Control Systems Engineering (CSE) Study Guide

Seventh Edition

Control Systems Engineering (CSE) Study Guide

for the Professional Engineering (PE) Licensing Examination

Seventh Edition

Notice

The information presented in this publication is for the general education of the reader. Because neither the author nor the publisher has any control over the use of the information by the reader, both the author and the publisher disclaim any and all liability of any kind arising out of such use. The reader is expected to exercise sound professional judgment in using any of the information presented in a particular application.

Additionally, neither the author nor the publisher has investigated or considered the effect of any patents on the ability of the reader to use any of the information in a particular application. The reader is responsible for reviewing any possible patents that may affect any particular use of the information presented.

Any references to commercial products in the work are cited as examples only. Neither the author nor the publisher endorses any referenced commercial product. Any trademarks or tradenames referenced in this publication, even without specific indication thereof, belong to the respective owner of the mark or name and are protected by law. Neither the author nor the publisher make any representation regarding the availability of any referenced commercial product at any time. The manufacturer's instructions on the use of any commercial product must be followed at all times, even if in conflict with the information in this publication.

Contents

Preface

Control Systems Engineering (CSE) was recognized as an engineering practice specialty by a vote of the National Council of Examiners for Engineering and Surveying (NCEES) at its annual meeting in Nashua, New Hampshire, on August 10, 1991. Recognition meant that exams would be developed by NCEES, with help from a professional society, and offered for use by boards for engineering licensing in the United States.

NCEES recognition was followed a request by control systems engineers to the Texas Board of Registration for Professional Engineers that it accept control system engineers for licensing in that state. The Texas Board, in turn, asked NCEES to provide an approved exam. Meetings with control system engineers, mostly Instrument Society of America (now the International Society of Automation[1]—ISA) members, showed NCEES committee and staff members that this new discipline satisfied their criteria for recognition: existence of an Accreditation Board for Engineering and Technology—accredited programs in the field; an effect on public health, safety, and welfare; and a need on the part of the profession; sufficient numbers to justify preparation of an exam; and availability of a professional society willing to support an exam program. An exam committee was required to submit a 2-year supply of acceptable exam problems.

As the sponsoring organization for the NCEES Control Systems Engineering PE exam, ISA cooperates with the CSE Committee in developing and updating study materials in support of its exam preparation courses.

CSE was first recognized in California in the 1970s. A limited period of licensing without examination (grandfathering) occurred in 1975–1976, and an exam was administered starting in 1978. The California exam did not meet NCEES standards and was not accepted by

---

[1] Refer to this webpage for more information on the history of ISA: https://www.isa.org/templates/two-column. aspx?pageid=111065.

other states for registration by reciprocity. This changed with the development of an NCEES-approved exam, administered for the first time in October 1992.

By April 2010, 46 member boards had agreed to offer the CSE exam. A few boards asked for a showing of need and interest in their state before they would recognize CSE. According to a recent poll, 52 of the 70 NCEES member boards offer the CSE exam. Three boards responded that they currently do not offer the NCEES CSE exam: Guam, Hawaii, and Rhode Island.

General Information

Introduction

The CSE exam is developed by the National Council of Examiners for Engineering and Surveying (NCEES). The membership of NCEES is composed of the boards of registration in 55 US jurisdictions (50 states, 4 territories, and the District of Columbia). NCEES is chartered to provide these boards with uniform, knowledge-based specifications for each recognized PE exam and to provide exams based on those specifications to measure the competency of candidates to practice engineering.

This study guide is published by the International Society of Automation (ISA) to assist candidates who are preparing for the Principles and Practices of Engineering exam in Control Systems Engineering (CSE) for the PE. This study guide is based on the specified areas of knowledge under the current control system engineer specification, as accepted by the jurisdictions.

To develop reliable and valid exams, NCEES employs procedures using the guidelines established in the *Standards for Educational and Psychological Testing* (1985), published by the American Psychological Association. These procedures are intended to maximize the fairness and quality of the exams. The procedures require the involvement of experienced testing specialists having the necessary expertise to develop exams using current testing techniques.

The exams are the result of careful preparation by committees comprised of PEs from throughout the United States. These engineers use their expertise to develop exam questions in accordance with the current specification. By utilizing the expertise of engineers with different backgrounds, such as private consulting, government, industry, and education, NCEES prepares exams that measure the competency of a candidate in multiple facets of control systems engineering.

Licensing Requirements

Eligibility

Eligibility varies from jurisdiction to jurisdiction. Access the NCEES website (NCEES.org) for the most current information.

Exam Schedule

Refer to the NCEES website for the most current schedule and registration information: https://ncees.org/engineering/pe/control-systems/.

Application Procedures and Deadlines

Requirements and fees vary from jurisdiction to jurisdiction. Access the NCEES website (NCEES.org) for the most current information.

Exam Description

Exam Format

Beginning in 2022, the exam is computer-based testing (CBT) and references are supplied by NCEES.

Exam Content

The subject areas of the CSE exam are described in the current exam specification located on the NCEES website.

Exam Development History

In 1991, ISA retained a contractor to conduct a comprehensive professional activities and requirements study of the CSE discipline. Questionnaires were sent to 3200 PEs practicing as control system engineers; approximately 800 replies were received. Based on their responses, a specification was developed for an exam to measure critical aspects of the CSE profession. The first exam was administered in October 1992. The exam specification was modified slightly when the all-multiple-choice format was initiated in 1998.

Similar studies are performed for all the disciplines in which NCEES provides exams. The studies are repeated periodically to reflect changes in technology. The exam specifications are updated on a regular basis as required by the jurisdictions.

Exam Procedures and Instructions

For the must current exam procedures and instructions, refer to the NCEES Examinee Guide located at https://ncees.org/exams/examinee-guide/.

Tips for Taking PE Exams

- Advance study, either individual or in an organized review course, is generally helpful in preparing for a PE exam. Surveys show that the principle of diminishing returns sets in after 40–100 hours of preparation.

- Last-minute cramming is probably not helpful. A good night's sleep is advised before the exam.

- Tape recorders, cameras, cell phones, walkie-talkie radios, and other communication devices are prohibited in exam rooms.

- Plan to arrive early at the exam site, allowing for delays in travel and parking.

- It is important to review the NCEES Examinee Guide (https://ncees.org/exams/examinee-guide/) so that you know the rules before taking the exam.

References for the CSE Exam

The following pages list reference books and standards that may be useful in preparing for the PE exam in CSE. The list has been organized into the same topic areas as the exam. Where possible, several books have been listed for each topic, and excerpts from the tables of contents are included to assist candidates in comparing these books with other similar references.

It is *not* suggested that candidates should be familiar with or own all the following books, because there are substantial overlaps in coverage of the exam content in the listed books. Instead, candidates should review these books and other similar books, select a limited number of references covering the major areas of the CSE exam, and study the selected references to learn where particular topics are covered.

Some of the listed references may be out of print or unavailable. However, their coverage of basic principles may still be valid and useful. Some older references may have been replaced by newer editions.

ISA offers a wide range of books, standards, electronic media products, and training courses on control systems engineering, some of which are referenced in this section. For more information, please visit the ISA website, www.isa.org, or contact ISA Customer Support at (919) 549-8411, fax (919) 549-8288.

No representation is made or intended that mastery of the content of the listed references is sufficient to assure that one will pass the CSE exam.

General References

This list is an example of the references that candidates have used to prepare for the exam. It provides a good starting point for developing a study reference list. The material that you use should be based on your background, references that you are familiar with, and your jurisdiction (state, territory, or district) guidelines and limitations. Make sure that you have the most current edition of any resource that you use.

ISBN	Title	Author/Editor/Publisher
978-1-937560-47-8	*Control Systems Engineer Technical Reference Handbook*	Chuck Cornell, PE, CAP, PMP ISA, 2012
978-1-945541-54-4	*Safety Instrumented Systems: A Life-Cycle Approach*	Paul Gruhn, PE, CFSE, and Simon Lucchini, CFGE, MIEAust, CPEng ISA, 2018
978-1-60650-124-5	*Advanced Regulatory Control: Applications and Techniques*	David W. Spitzer Momentum Press, 2009
0-87201-382-0	*Applied Instrumentation in the Process Industries, Volume I: A Survey,* Second Edition	W. G. Andrew and H. B. Williams Gulf Professional Publishing, 1979
0-87201-047-3	*Applied Instrumentation in the Process Industries, Volume III: Engineering Data and Resource Material,* Third Edition	W. G. Andrew and H. B. Williams Gulf Professional Publishing, 1993
0-8019-6766-x	*Instrumentation for Process Measurement and Control,* Third Edition	Norman A. Anderson Taylor & Francis, 1998
978-0-07150-000-5	*Process/Industrial Instruments and Controls Handbook,* Fifth Edition	Gregory K. McMillan and Douglas M. Considine McGraw Hill Professional, 1999
978-0-47186-608-4	*On-Line Process Analyzers*	Gary D. Nichols Wiley-Interscience, 1988
978-1-945541-04-9	*Industrial Ethernet: How to Plan, Install, and Maintain TCP/IP Ethernet Networks,* Third Edition	John S. Rinaldi and Perry S. Marshall ISA, 2017
No ISBN Number	*Flow of Fluids Through Valves, Fittings, and Pipe, Technical Paper No. 410*	Crane, 2018
No ISBN Number	*Flow Meter Engineering Handbook,* Fourth Edition	C. F. Cusick, Honeywell Minneapolis-Honeywell Regulator Co., 1968
952-9773-12-9	*Flow Control Manual,* Sixth Edition	Jari Kirmanen, et al. Metso Automation Inc., 2011
No ISBN Number	*Pumps Engineering Data*	Gorman-Rupp Pumps 2008–2012 https://www.grpumps.com/files/ AV-06196.pdf
978-0-9766259-4-0	*Programmable Logic Controllers: An Emphasis on Design and Application,* Third Edition	Kelvin T. Erickson Dogwood Vally Press, LLC, 2016
No ISBN Number	*Control Valve Handbook,* Fifth Edition	Fisher Controls International, 2019 https://www.emerson.com/ documents/automation/manuals-guides-control-valve-handbook-fisher-en-3661206.pdf
CHD-LCO2150	*Cameron Hydraulic Data,* 19th Edition	Flowserve Corporation, 2010
978-1-945541-38-4	*The Condensed Handbook of Measurement and Control,* Fourth Edition	N. E. Battikha, PE ISA, 2018
978-0-07042-366-4	*Flow Measurement Engineering Handbook,* Third Edition	Richard W. Miller McGraw-Hill, 1996

ISBN	Title	Author/Editor/Publisher
No ISBN Number	*Consolidated* Valve Sizing and Selection* Rev. E	Baker Hughes, 2022 chrome-extension:// efaidnbmnnnibpcajpcglclefindmkaj/ https://dam.bakerhughes.com/ m/53e77b4e98239f8f/original/ CN-Valve-Sizing-Tech-Spec-GEA31796C-English-pdf.pdf
978-0-08097-470-5	*The Safety Relief Valve Handbook* First Edition	Marc Hellemans Butterworth-Heinemann, 2009
978-1-55617-859-7	*Variable Speed Drives: Principles and Applications for Energy Cost Savings,* Third Edition	David W. Spitzer ISA, 2004
978-1-55617-800-9	*Motors & Drives: A Practical Technology Guide,* Second Edition	David Polka ISA, 2020
978-1-941546-56-7	*Control Systems Engineering Exam Reference Manual: A Practical Study Guide,* Fourth Edition	Byron Lewis, CSE, PE ISA, 2020
978-0-84931-083-6	*Instrument Engineers' Handbook, Fourth Edition, Volume One: Process Measurement and Analysis*	Béla G. Liptak CRC Press, 2003
978-0-84931-081-2	*Instrument Engineers' Handbook, Fourth Edition, Volume Two: Process Control and Optimization*	Béla G. Liptak CRC Press, 2005

Codes and Standards

The following list contains codes and standards relevant to the practice of CSE. The source, number, and title of the codes or standards are given. Candidates should study the most current versions.

Because there are so many applicable codes and standards, it is not expected that CSEs will memorize all their provisions.

International Society of Automation (ISA)

P.O. Box 12277
Research Triangle Park, NC 27709
(919) 549-8411

- ANSI/ISA-5.1, *Instrumentation Symbols and Identification*

- ISA-TR5.1.01 - ISA-TR77.40.01, *Functional Diagram Usage*

- ISA-5.2, *Binary Logic Diagrams for Process Operations*

- ISA-5.3, *Graphic Symbols for Distributed Control/Shared Display Instrumentation, Logic, and Computer Systems*

- ISA-5.4, *Standard Instrument Loop Diagrams*

- ISA-5.5, *Graphic Symbols for Process Displays*

- ANSI/ISA-12.00.02, *Certificate Standard for AEx Equipment for Hazardous (Classified) Locations*

- ANSI/ISA-12.01.01, *Definitions and Information Pertaining to Electrical Equipment in Hazardous (Classified) Locations*

- ANSI/ISA-12.04.04, *Pressurized Enclosures*

- ISA-18.1, *Annunciator Sequences and Terminology*

- ISA-51.1, *Process Instrumentation Terminology*

- ANSI/ISA-75.01.01 (60534-2-1 MOD), *Industrial-Process Control Valves – Part 2-1: Flow Capacity – Sizing Equations for Fluid Flow Under Installed Conditions*

- ANSI/ISA-75.11.01, *Inherent Flow Characteristics and Rangeability of Control Valves*

- ANSI/ISA-61511-1 / IEC 61511-1 + AMD1 CSV, *Functional Safety – Safety Instrumented Systems for the Process Industry Sector – Part 1: Framework, Definitions, System, Hardware and Application Programming Requirements (IEC 61511-1 + AMD1 CSV, IDT)*

- ANSI/ISA-61511-2 / IEC 61511-2, *Functional Safety – Safety Instrumented Systems for the Process Industry Sector – Part 2: Guidelines for the Application of IEC 61511-1 (IEC 61511-2, IDT)*

- ANSI/ISA-61511-3 / IEC 61511-3, *Functional Safety – Safety Instrumented Systems for the Process Industry Sector – Part 3: Guidance for the Determination of the Required Safety Integrity Levels (IEC 61511-3, IDT)*

- ISA-RP60.3, *Human Engineering for Control Centers*

- ISA-TR99.00.01, *Security Technologies for Industrial Automation and Control Systems*

- ANSI/ISA-60079-0 (12.00.01), *Explosive Atmospheres – Part 0: Equipment – General Requirements*

- ANSI/ISA-61511-3 /IEC 61511-3, *Functional Safety – Safety Instrumented Systems for the Process Industry Sector – Part 3: Guidance for the Determination of the Required Safety Integrity Levels* (IEC 61511-3, IDT)

- ANSI/ISA 62443-4-1, *Security for Industrial Automation and Control Systems – Part 4-1: Product Security Development Life-Cycle Requirements*

American National Standards Institute (ANSI)

25 W 43rd Street
New York, NY 10036
(212) 642-4900

- ANSI/ASME PTC 19.1, *Test Uncertainty*

- ANSI/ASME PTC 19.2, *Pressure Measurement Instruments and Apparatus*

- ASME PTC 19.3 TW, *Thermowells*

- ASME PTC 19.22, *Data Acquisition Systems*

American Petroleum Institute (API)

1220 L Street, NW
Washington, DC 20005-4070
(202) 682-8000

- ANSI/API Spec 5L, *Specification for Line Pipe*

- API RP 500, *Recommended Practice for Classification of Locations for Electrical Installations at Petroleum Facilities Classified as Class I, Division I and Division 2*

- ANSI/API RP 505, *Recommended Practice for Classification of Locations for Electrical Installations at Petroleum Facilities Classified as Class 1, Zone 0, and Zone 2*

- API STD 520, *Sizing, Selection, and Installation of Pressure-Relieving Devices in Refineries, Part I - Sizing and Selection*

- API STD 521, *Pressure-Relieving and Depressuring Systems*

American Society of Mechanical Engineers (ASME)

Three Park Avenue
New York, NY 10016-5990
(800) 843-2763 (US/Canada); (973) 882-1170 (outside North America); 001-800-843-2763 (Mexico)

- ASME Section I, *ASME Boiler and Pressure Vessel Code, Section I: Rules for Construction of Power Boilers*

- ASME Section IV, *ASME Boiler and Pressure Vessel Code, Section IV: Rules for Construction of Heating Boilers*

- ASME Section VIII SET, *ASME Boiler and Pressure Vessel Code, Section VIII: Rules for Construction of Pressure Vessels*

- ANSI/ASME B16.5, *Pipe Flanges and Flanged Fittings: NPS 1/2 through NPS 24 Metric/Inch Standard*

- ASME B31.9, *Building Services Piping*

- ASME/ANSI MFC Series

  o MFC-7, *Measurement of Gas Flow by Means of Critical Flow Venturis and Critical Flow Nozzles*

  o MFC-7, *Measurement of Gas Flow by Means of Critical Flow Venturis and Critical Flow Nozzles*

  o MFC-21.1, *Measurement of Gas Flow by Means of Capillary Tube Thermal Mass Flowmeters and Mass Flow Controllers*

  o (MFC-1, *Glossary of Terms Used in the Measurement of Fluid Flow in Pipes*)

- o MFC-16, *Measurement of Liquid Flow in Closed Conduits with Electromagnetic Flowmeters*

- o MFC-6, *Measurement of Fluid Flow in Pipes Using Vortex Flowmeters*

- o MFC-5.3, *Measurement of Liquid Flow in Closed Conduits Using Doppler Ultrasonic Flowmeters*

- o MFC-26, *Measurement of Gas Flow by Bellmouth Inlet Flowmeters*

- o MFC-5.1, *Measurement of Liquid Flow in Closed Conduits Using Transit-Time Ultrasonic Flowmeters*

- o MFC-21.2, *Measurement of Fluid Flow by Means of Thermal Dispersion Mass Flowmeters*

- o MFC-19G, *Wet Gas Flowmetering Guideline* (Technical Report)

- o MFC-22, *Measurement of Liquid by Turbine Flowmeters*

- o MFC-11, *Measurement of Fluid Flow by Means of Coriolis Mass Flowmeters*

- o MFC-13M, *Measurement of Fluid Flow in Closed Conduits: Tracer Methods*

- o MFC-12M, *Measurement of Fluid Flow in Closed Conduits Using Multiport Averaging Pitot Primary Elements*

- o MFC-3M, *Measurement of Fluid Flow in Pipes Using Orifice, Nozzle, and Venturi*

- o MFC-14M, *Measurement of Fluid Flow Using Small Bore Precision Orifice Meters*

- o MFC-10M, *Method for Establishing Installation Effects on Flow Meters*

- o MFC-18M, *Measurement of Fluid Flow using Variable Area Meters*

- o MFC-8M, *Fluid Flow in Closed Conduits: Connections for Pressure Signal Transmissions Between Primary & Secondary Devices*

- o MFC-2M, *Measurement Uncertainty for Fluid Flow in Closed Conduits*

- o MFC-4M, *Measurement of Gas Flow by Turbine Meters*

- o MFC-3M, *Measurement Of Fluid Flow In Pipes Using Orifice, Nozzle and Venturi*

- o MFC-9M, *Measurement of Liquid Flow in Closed Conduits by Weighing Method*

International Electrotechnical Commission (IEC)

3, rue de Varembé
P.O. Box 131
CH - 1211 Geneva 20 – Switzerland
41 22 919 02 11

- • IEC 60529, *Degrees of Protection Provided by Enclosures (IP Code)*

- • IEC 60079-7, *Explosive Atmospheres – Part 7: Equipment Protection by Increased Safety "e"*

- • IEC 60079-10-1, *Explosive Atmospheres – Part 10-1: Classification of Areas - Explosive Gas Atmospheres*

- IEC 60079-10-2, *Explosive Atmospheres – Part 10-2: Classification of Areas - Combustible Dust Atmospheres*

- IEC 60079-11, *Explosive Atmospheres – Part 11: Equipment Protection by Intrinsic Safety "i"*

- IEC 62443-4-1, *Security for industrial automation and control systems – Part 4-1: Secure product development lifecycle requirements*

National Fire Protection Association (NFPA)

1 Batterymarch Park
Quincy, MA 02269-9101
(617) 770-3000

- No. 70, *National Electrical Code Handbook*

- No. 72, *National Fire Alarm and Signaling Code*

- No. 85, *Boiler and Combustion Systems Hazards Code*

- No. 496, *Standard for Purged and Pressurized Enclosures for Electrical Equipment*

- No. 497, *Recommended Practice for the Classification of Flammable Liquids, Gases, or Vapors and of Hazardous (Classified) Locations for Electrical Installations in Chemical Process Areas*

National Electrical Manufacturers Association (NEMA)

Suite 1847
1300 N 17th Street
Rosslyn, VA 22209
(703) 841-3200

- ICS 6, *Enclosures*

Institute of Electrical and Electronics Engineers (IEEE)

445 Hoes Lane
Piscataway, NJ 08854-1331
(732) 981-0060

- S 315A, *Supplement to Graphic Symbols for Electrical and Electronics Diagrams*

- ANSI/IEEE 488, *IEEE Standard Digital Interface for Programmable Instrumentation*

- 802.1-Q, *IEEE Standard for Local and Metropolitan Area Networks – Bridges and Bridged Networks*

- 991, *IEEE Standard for Logic Circuit Diagrams*

Occupational Safety & Health Administration (OSHA)

200 Constitution Avenue NW
Washington, DC 20210
(202) 693-2000

- 1910 Occupational Safety and Health Standards, *Subpart H, Hazardous Materials 1910.119, Process Safety Management of Highly Hazardous Chemicals*

Readers of this book are urged to submit information (name of author(s), title, edition, publisher's name and address, date of publication, and description of contents) for other pertinent references, codes, or standards to the address below so they can be included in future editions of the *CSE Study Guide*.

ISA
P.O. Box 12277
Research Triangle Park, NC 27709
Phone: (919) 549-8411 Fax: (919) 549-8288

Appendix A
CSE Exam Specification

This appendix contains the official NCEES exam specification, in the form distributed by its member boards. There are five major areas of activity for control systems engineers, identified as I through V, with a descriptive title and the approximate number of questions on the exam devoted to that area. For each area of activity, associated subareas of knowledge are indicated.

NCEES Principles and Practice of Engineering Exam

Control Systems Exam Specifications

Effective Beginning October 1, 2022

- The exam topics have not changed since October 2019 when they were originally published.

- The exam is computer-based. It is closed book with an electronic reference. Design standards applicable to the PE Control Systems exam are shown on the last page.

- Examinees have 9.5 hours to complete the exam, which contains 85 questions. The 9.5-hour time includes a tutorial and an optional scheduled break. Examinee works all questions.

- The exam uses both the International System of Units (SI) and the US Customary System (USCS).

- The exam is developed with questions that will require a variety of approaches and methodologies, including design, analysis, and application.

- The knowledge areas specified as examples of kinds of knowledge are not exclusive or exhaustive categories.

1. Measurement

A. Sensors
1. Sensor technologies applicable to general measurement (e.g., flow, pressure, level, temperature, counters, motion)
2. Sensor technologies applicable to general analytical instruments and sampling systems (e.g., pH, ORP, density, O_2, conductivity, effects of sampling systems, GC)
3. Sensor technologies applicable to fire and gas detection
4. Sensor technologies applicable to machinery monitoring and protection (e.g., vibration, bearing temperature, lube oil pressures, thrust, speed)
5. Sensor characteristics (e.g., rangeability, accuracy and precision, temperature effects, response times, reliability, repeatability, maintenance, calibration)
6. Sensor selection (e.g., plugging service, process severity, environmental effects and constraints, costs)
7. Material compatibility
8. Installation details (e.g., process, pneumatic, electrical, location, maintenance, calibration)

B. Flow, Level, and Pressure Calculations
1. Flow (e.g., element sizing, pressure-temperature compensation, mass/volume)
2. Level
3. Pressure drop

C. General Calculations
1. Unit conversions
2. Velocity
3. Square root extraction and interpolation
4. Variables involved in wake frequency calculations (e.g., thermowell length/diameter, velocity, natural frequency, wake frequency)

2. Control Systems   **17–27**

A. Drawings
  1. Drawings (e.g., process flow diagrams, P&IDs, loop diagrams, ladder diagrams, logic drawings, cause and effects drawings, electrical drawings, schematics, wiring diagrams)

B. Theory
  1. Basic control of processes (e.g., pumps, compression, combustion, evaporation, distillation, hydraulics, reaction, dehydration, heat exchangers, crystallization, filtration, refrigeration, fluidization)
  2. Process dynamics (e.g., loop response, pressure-volume-temperature relationships, simulations)
  3. Basic control (e.g., regulatory control, feedback, feedforward, cascade, ratio, PID, split-range, gap control)
  4. Discrete control (e.g., relay logic, Boolean algebra, aliasing)
  5. Sequential control (e.g., batch, assembly, conveying, CNC, state machine, sequential function chart)

C. Implementation
  1. HMI (e.g., graphics, alarm management, trending, historical data, operator panels)
  2. Equipment layout (e.g., human factors engineering, physical control room arrangement, panel layout)
  3. Limited variability programming languages for DCS and PLC (e.g., IEC 61131-3 languages/ladder diagrams, function blocks, sequential function charts, structured text, instruction list)
  4. System design comparisons and compatibilities (e.g., advantages and disadvantages of system architecture, distributed architecture, remote I/O, buses, wireless)
  5. Installation requirements (e.g., shielding, constructability, I/O termination, environmental, heat load calculations, power load requirements, purging, lighting, maintainability)
  6. System testing (e.g., factory acceptance test, integrated system test, site acceptance test)
  7. Commissioning (e.g., performance tuning, loop checkout)
  8. Performance evaluation (e.g., troubleshooting, root cause failure analysis and correction)

D. Security of Industrial Automation and Control Systems
  1. Security (e.g., physical, cyber, network, firewalls, routers, switches, protocols, hubs, segregation, access controls)
  2. Security life cycle (e.g., assessment, controls, audit, management of change)
  3. Requirements for a security management system
  4. Security risk assessment and system design
  5. Product development and requirements
  6. Verification of security levels (e.g., level 1, level 2)

3. Final Control Elements   14–23
  A. Valves
     1. Types (e.g., globe, ball, butterfly)
     2. Trim characteristics (e.g., linear, low noise, equal percentage, seat leakage class)
     3. Calculation (e.g., sizing, split range, noise, actuator, response time, pressure drop, air/gas consumption)
     4. Selection of motive power and failure mode (e.g., hydraulic, pneumatic, electric, spring)
     5. Applications of fluid dynamics (e.g., cavitation, flashing, choked flow, Joule-Thompson effects, two-phase)
     6. Material selection based on process characteristics (e.g., erosion, corrosion, plug, extreme pressure, temperature, material compatibility)
     7. Accessories (e.g., limit switches, solenoid valves, positioners, transducers, air regulators, servo amp, boosters, quick exhaust)
     8. Environmental constraints (e.g., fugitive emissions, packing, special sealing, fire rating)
     9. Installation practices (e.g., vertical, horizontal, bypasses, location, flow direction)
  B. Pressure Relieving Devices
     1. Pressure relieving valve types (e.g., conventional spring, balanced bellows, pilot operated)
     2. Pressure relieving valve characteristics (e.g., modulating, pop action)
     3. Pressure relieving valve calculations (e.g., sizing considering inlet pressure drop, back pressure, multiple valves)
     4. Material selection based on process characteristics
     5. Pressure relieving valve installation practices (e.g., linking valves, sparing the valves, accessibility for testing, car sealing inlet valves, piping installation, combination devices)
     6. Rupture discs and buckling pin valves (e.g., types, characteristics, application, calculations)
  C. Motor Controls
     1. Types (e.g., motor starters, variable-speed drives)
     2. Applications (e.g., speed control, soft starters, motor-operated valve actuators)
     3. Calculations (e.g., sizing, tuning, location)
     4. Accessories (e.g., encoders, positioners, relays, limit switches)
  D. Other Final Control Elements
     1. Motion (e.g., damper controls, types, orientation, actuators, servos, encoders)
     2. Solenoid valves (e.g., types, sizing)
     3. On-off devices/relays (e.g., types, applications, energize and de-energize to trip)
     4. Self-regulating devices (e.g., types, sizing, pressure, temperature, level, and flow regulators)

4. Signals, Transmission, and Networking   **11–18**
  A. Signals
    1. Pneumatic, electronic, optical, hydraulic, digital, analog, buses, wireless, thermocouple
    2. Transducers (e.g., analog/digital [A/D], digital/analog [D/A], current/pneumatic [I/P] conversion, current/current [I/I], splitters, filters)
    3. Hazardous area classification and instrument installation techniques (e.g., intrinsically safe [IS] barriers, cabinet purges, non-incendive)
    4. Grounding, shielding, segregation, electromagnetic interference
    5. Basic signal circuit design (e.g., two-wire, four-wire, isolated outputs, loop powering, buses)
    6. Circuit calculations (voltage, current, impedance, power)
    7. Unit conversion calculations
  B. Transmission
    1. Different communication systems architecture and protocols (e.g., fiber optics, coaxial cable, wireless, paired conductors, buses, transmission control protocol/internet protocol [TCP/IP], OPC)
    2. Distance considerations versus transmission medium (e.g., data rates, sample rates)
  C. Networking
    1. Routers, bridges, switches, firewalls, gateways, network loading, error checking, bandwidth, crosstalk, parity, hubs

5. Safety Systems   **11–19**
  A. Documentation
    1. Basic documentation required (e.g., process hazards analysis, safety requirements specification [SRS], logic diagrams/narratives, test procedures, SIL selection report, SIL verification report, safety life-cycle plan)
  B. Theory
    1. Reliability and availability (e.g., bathtub curve, failure rates types, voting, proof test intervals, common cause and diversity)
    2. SIL selection (e.g., safety layer matrix, risk graph, LOPA)
  C. Implementation
    1. Safety system design (e.g., SRS, I/O assignments, redundancy, segregation, logic design, failure direction)
    2. SIL verification calculations (e.g., failure rates types, voting, proof test intervals, common cause and diversity)
    3. Installation, commissioning, and validation (e.g., methods, procedures, test records)
  D. Safety Life-Cycle Management
    1. Modifications (e.g., management of change, scope of change, impact of change, documentation)
    2. Operations and maintenance (e.g., methods, procedures, test records, partial stroke testing, demand tracking, bypass and override management, failure analysis, validation of design assumptions)

In addition to the *PE Control Systems Reference Handbook*, the following codes and standards will be supplied in the exam as searchable, electronic pdf files with links for easy navigation.

Solutions to exam questions that reference a standard of practice are scored based on this list and the revision year shown. Solutions based on other standards will not receive credit.

NCEES does not sell design standards or printed copies of the NCEES handbook. The NCEES handbook is accessible from your MyNCEES account. Design standards are available through the publisher or a bookseller.

ABBREVIATION	DESIGN STANDARD TITLE
ANSI/ISA-5.1	*Instrumentation Symbols and Identification*, 2009, International Society of Automation, Research Triangle Park, NC, www.isa.org.
ISA/IEC 61511	*Functional Safety – Safety Instrumented Systems for the Process Industry Sector – Part 1: Framework, Definitions, System, Hardware and Application Programming Requirements*, 2018, International Society of Automation, Research Triangle Park, NC, www.isa.org.

Appendix B
Sample Questions

This is the seventh study guide published for the CSE exam. The following questions have been assembled from various sources, including past exams in CSE and other disciplines; some questions have been written specifically for this publication. Whereas there are 85 questions on the CBT CSE exam, and the time frame is 9.5 hours, this practice exam consists of 80 questions, and it should be completed within 8 hours.

The following questions are intended to illustrate the types of questions that may be encountered in the CSE exam. Questions in this study guide will *not* appear in exams.

The sample questions conform to the NCEES exam specification in having the prescribed number of questions for each listed topic. The question numbering format on the CBT exam may not be the same as the format on the previous paper-style exam; the numbering for these sample questions is in the paper-style exam format. Half of the questions are numbered 101 to 140, representing the first part of the exam before the break; the others are numbered 501 to 540, representing the second part of the exam after the break. To the extent possible, questions on a particular topic are divided equally between the two parts of the exam.

Because the following questions resemble an actual exam, candidates wishing to evaluate their performance under test conditions can do so. Select a time and place where you will not be interrupted, assemble the materials you would take to the actual exam, and allow yourself up to 4 hours for each half of the exam. To create an answer form, use a sheet from a pad of lined writing paper; make four columns, numbered 101–120, 121–140, 501–520, and 521–540.

Answers and/or solutions are given separately in Appendix C so that users of the study guide can read the questions and record their responses without simultaneous exposure to the answers. The answers in Appendix C are identified by topic areas, so you can easily see which areas you should focus on when preparing to take the actual exam.

Candidates should understand that any one CSE exam can cover only some of the areas of activity and knowledge listed in the exam specification. Thus, the following questions do not necessarily deal with all the possible subareas of CSE activity or knowledge that may appear in CSE exams.

Likewise, while an effort was made to match the level of difficulty of actual exams, the match may not be exact. Variations in difficulty from exam to exam are considered in setting the passing scores. For this reason, candidates should view their ability to answer the sample questions as an indication of where to focus their preparatory efforts, not as a prediction of their success on the actual exam.

First Session

101. The flow rate of a clean, low-viscosity liquid is to be measured as the process input to a flow control loop. The loop has a 4:1 turndown ratio, and the accuracy requirement is 2%. The flow rate is best measured using a(n):

    (A) Thermal

    (B) Positive displacement meter

    (C) Pitot tube

    (D) Orifice plate

102. Online measurement of the 90% point of a gasoline-blending component is best done using a:

    (A) Liquid chromatograph

    (B) Mass spectrometer

    (C) Boiling point analyzer

    (D) Infrared analyzer

103. A differential pressure transmitter (LT-100) is used to monitor the level in a horizontal storage vessel (V-100) that contains hot water. The vessel is manually filled and drained by valves V-1 and V-2. The pressure taps on the vessel for LT-100 are 60 in apart, and the transmitter is mounted 10 in below the bottom process tap. The transmitter's process tubing is routed 10 ft horizontally before dropping vertically to the transmitter.

    The following data apply:

    Ambient temperature: 80°F (always)

    Normal operating pressure: 100 psig (PIC-101)

    Normal operating temperature: 280°F (TIC-102)

Seal fluid for reference leg of level transmitter:

SG = 0.8 at 60°F

SG = 0.62 at 80°F

SG = 0.32 at 280°F

Boiling point = 520°F at 0 psig

The other leg of the level transmitter contains process fluid.

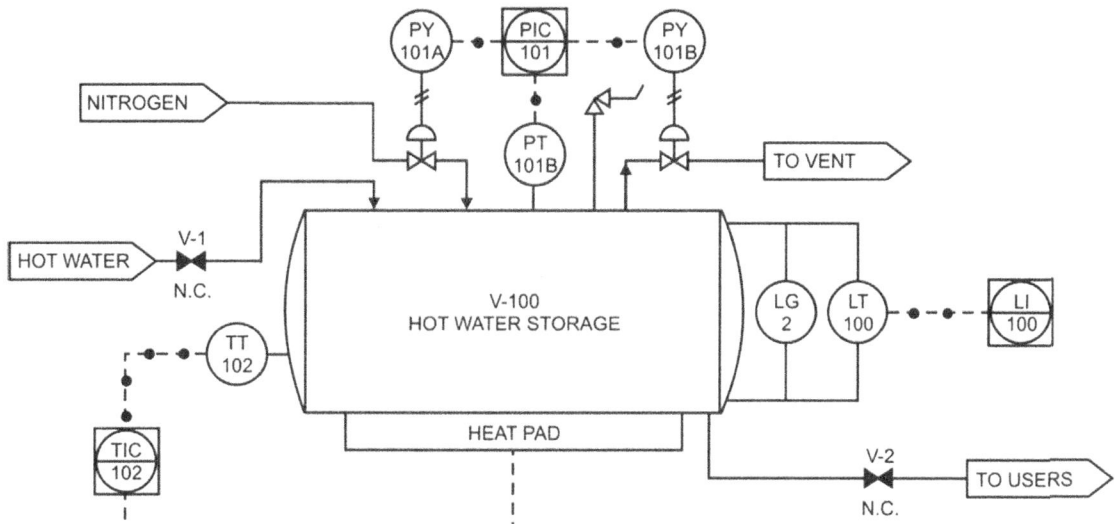

The span of the transmitter, in inches of water, is:

(A) 19.2   (B) 37.2

(C) 55.7   (D) 60.0

104. You must provide a tank gauging system for a liquefied natural gas (LNG) tank. LNG is a cryogenic liquid and is stored at near atmospheric pressure. Under certain conditions, the LNG can stratify and potentially lead to unsafe conditions. The best choice to provide reliable, safe operation of the LNG tank would be:

(A) Radar type with independent multipoint temperature indication that provides a temperature profile of the tank

(B) Differential pressure with multiple temperature indicators located at low liquid level around the diameter of the tank

(C) Servo type that provides liquid level plus temperature and density profiles

(D) Float type with an independent multipoint temperature for tank temperature profile and hydrostatic density indication

105. A liquid with an SG of 0.8 flows to a pump at 100 psig. The viscosity of the liquid is 3 cP, the molecular weight is 94, and the temperature is 15°C. The pump supplies 100 ft of head. Piping losses are insignificant. The pressure (psig) at the inlet to the control valve will be most nearly:

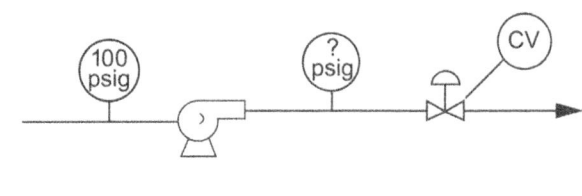

(A) 35   (B) 44

(C) 135   (D) 154

106. TK-1 has a capacity of 30,000 barrels (US and must be filled in 6 hours. FV-1 must be sized for a flow rate (gpm) of most nearly:

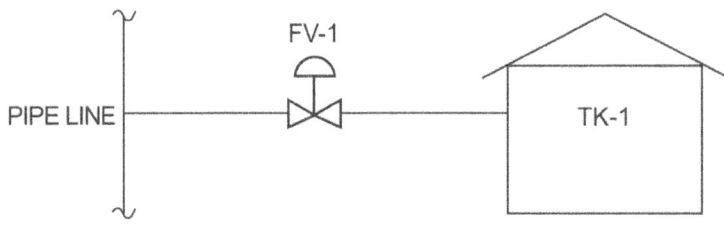

(A) 3500   (B) 5000

(C) 7143   (D) 21,000

107. Which of the following gas chromatograph (GC) detector types is most appropriate for measuring low concentrations of sulfur in hydrocarbon fuels?

(A) Thermal conductivity (TCD)

(B) Flame ionization (FID)

(C) Flame photometric (PFD)

(D) Electron capture (ECD)

108. Which pair of devices listed below are most likely to be hazardous gas/gas environment detectors?

(A) Catalytic bead, piezoelectric adsorption

(B) P_2O_5, infrared

(C) Catalytic bead, open path

(D) Electrochemical, thermal conductivity

109. A safety barrier limits the energy transfer to a hazardous location by limiting the maximum current (using a fuse) and shunting any high-voltage faults in the safe area to a safety ground (through Zener diodes).

Which of the following considerations does **not** apply to a shunt Zener diode barrier?

(A) Requires an intrinsically safe ground

(B) Can cause voltage drop problems

(C) Is expensive and difficult to use

(D) Fuse can blow during start up

110. A control loop of a DCS consists of an isolated 4–20 mA output and a grounded control element. A remote indicating device must be installed between the DCS and the control element. The indicating device has an input range of 1–5 V and is referenced to ground. Which of the following components (I/I is a current-to-current converter) is/ are required to give a full-scale reading of the indicating device without affecting the control element?

(A) 100 Ω resistor only

(B) I/I and 100 Ω resistor

(C) 250 Ω resistor only

(D) I/I and 250 Ω resistor

111. Which of the following statements about wireless (radio) data transmission systems for use in plant-wide data collection and control applications is false?

(A) Wireless systems can have unlimited numbers of data links.

(B) Wireless systems are easier to install than wired or cable systems.

(C) Wireless systems do not eliminate the need to deliver power to sensor and/or controller locations.

(D) Wireless systems can be made highly resistant to ambient electrical noise.

112. Which of the following practices is important in routing fiber-optic cable?

(A) Laying cable in trays with high-horsepower motor wiring should be avoided.

(B) Conduit fittings that require small-radius bends should be avoided.

(C) Overhead runs on messenger wires should be limited to 75 ft.

(D) Underground fiber-optic runs must be covered with concrete.

113. What is the principal advantage of a fieldbus installation over traditional 4–20 mA with HART protocol?

    (A) Lower-cost field devices

    (B) Device reaction time

    (C) Device diagnostic coverage

    (D) Lower cost field cabling

114. Which physical network topologies provide fault tolerance?

    I   Ring

    II   Star

    III Bus

    IV   Mesh

    (A) I and IV

    (B) I and II

    (C) II and IV

    (D) I, II, III, and IV

115. A control valve, originally supplied for gaseous service, is now being considered for a liquid service application.

    Original service conditions:

    Maximum capacity = 60,000 scfh

    ΔP for valve sizing = 50 psi

    Gas molecular weight = 44

    Inlet pressure = 300 psig

    Inlet temperature = 120°F

    New service conditions:

    ΔP for valve sizing = 10 psi

    Liquid SG = 0.81

    Inlet pressure = 240 psig

    The valve coefficient, C_v, for the original service conditions is approximately equal to:

    (A) 4.8   (B) 6.9

    (C) 8.5   (D) 10.4

116. Which of the following statements about control valve installation practices is false?

    (A) At least 10–20 pipe diameters of straight run inlet piping should be provided upstream of control valves.

    (B) At least 3–5 pipe diameters of straight run piping should be provided downstream of control valves.

    (C) Diaphragm-actuated control valves must be installed with the stem in a vertical (upward) position.

    (D) Valves must be installed with the flow arrow in the correct direction.

117. Which of the following statements about control-valve installation practices is false?

    (A) If the material is not hazardous, vent lines are not required for fluid trapped between stop valves.

    (B) Valves must be located where an operator can see the indicators or gauges for manual control.

    (C) Lines should be flushed or blown out before valve installation.

    (D) Higher flow velocities are allowed in gas lines than in liquid or steam lines.

118. A suggestion is made that a rising stem globe valve should be replaced with a same size rotary stem ball valve. Which of the following statements is correct?

    (A) The ball valve C_v would likely be too large for the application.

    (B) The piping must be changed because the ball valve will likely be longer than the globe valve.

    (C) Using a ball valve would not allow use of a diaphragm-type actuator.

    (D) Ball valves are not used for modulating control.

119. The applicable ASME (American Society of Mechanical Engineers) code limits the maximum superimposed constant back pressure on pressure relief devices to what value?

    (A) Set pressure including effect of static head and back pressure

    (B) Set pressure including effect of static head

    (C) Fluid critical pressure

    (D) 55% above set pressure

120. What is the maximum vessel pressure (in psig) allowed by ASME Code Section VIII when only PSV-1 is in service and relieving?

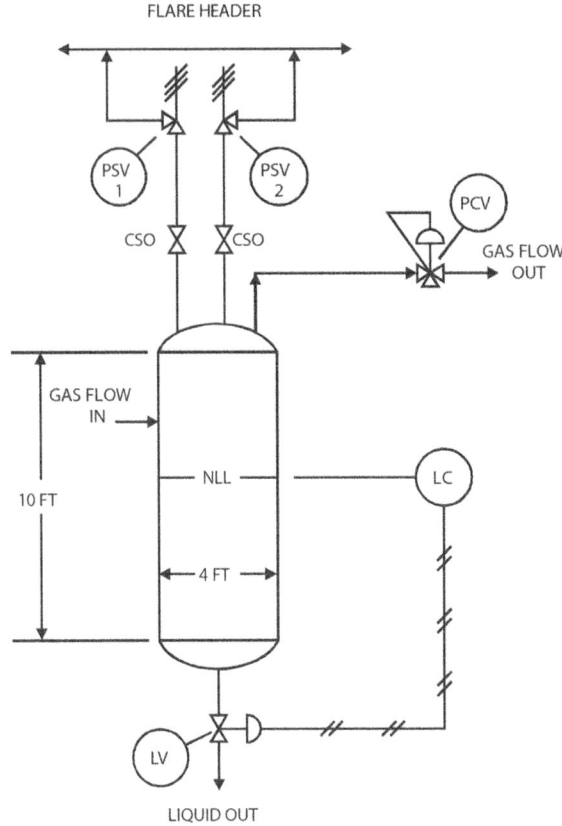

Vessel data:

    Maximum allowable working pressure (MAWP): 200 psig

    Dimensions: 4 ft OD, 10 ft cylinder, elliptical heads

    Normal liquid level: 3.5 ft

    Operating pressure: 150 psig

    Operating temperature: 100°F

Process data:

    Inlet gas flow: 80,000 lb/h

    Liquid flow: 60 gpm

    Molecular weight: 19

    SG: 0.590

(A) 200

(B) 210

(C) 220

(D) 230

121. What is the most acceptable, cost-effective method for protecting a safety relief valve in corrosive service?

   (A) Inert gas purge

   (B) Reverse-buckling rupture disk

   (C) Standard rupture disk

   (D) Wetted parts of noncorrosive materials

122. How would you insert a DCS RUN/STOP using a single DO into a motor control that uses a local momentary START/STOP where the local stop can shut down the motor?

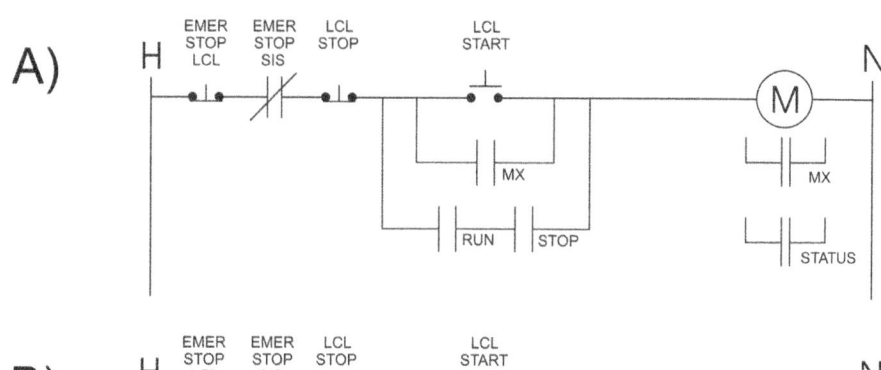

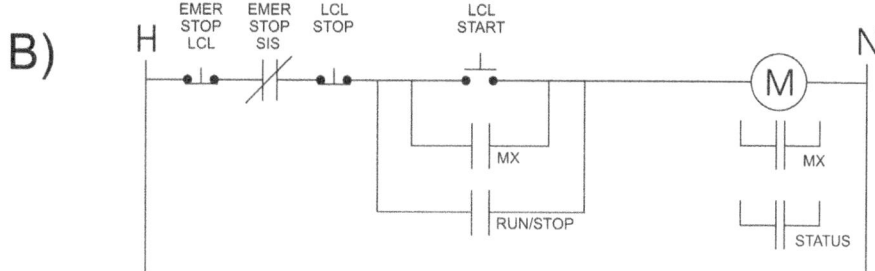

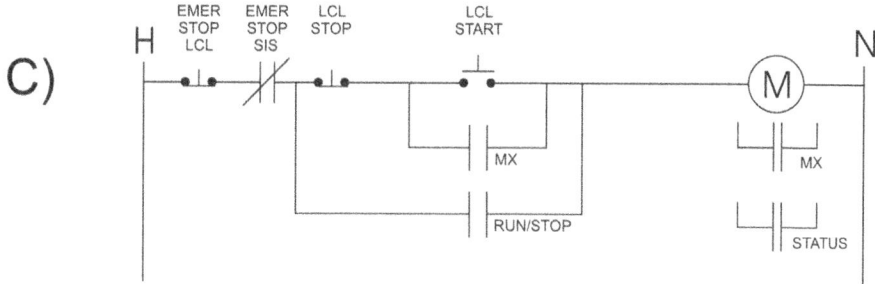

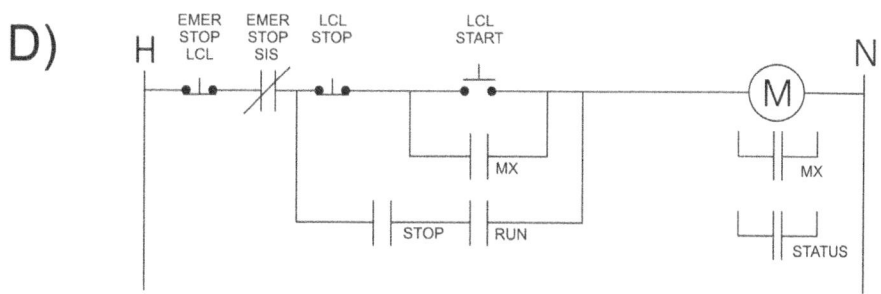

123. An HVAC blower has a shutdown from a building fire detection/protection panel for confirmed fire. Which of the following motor controls is the most correct?

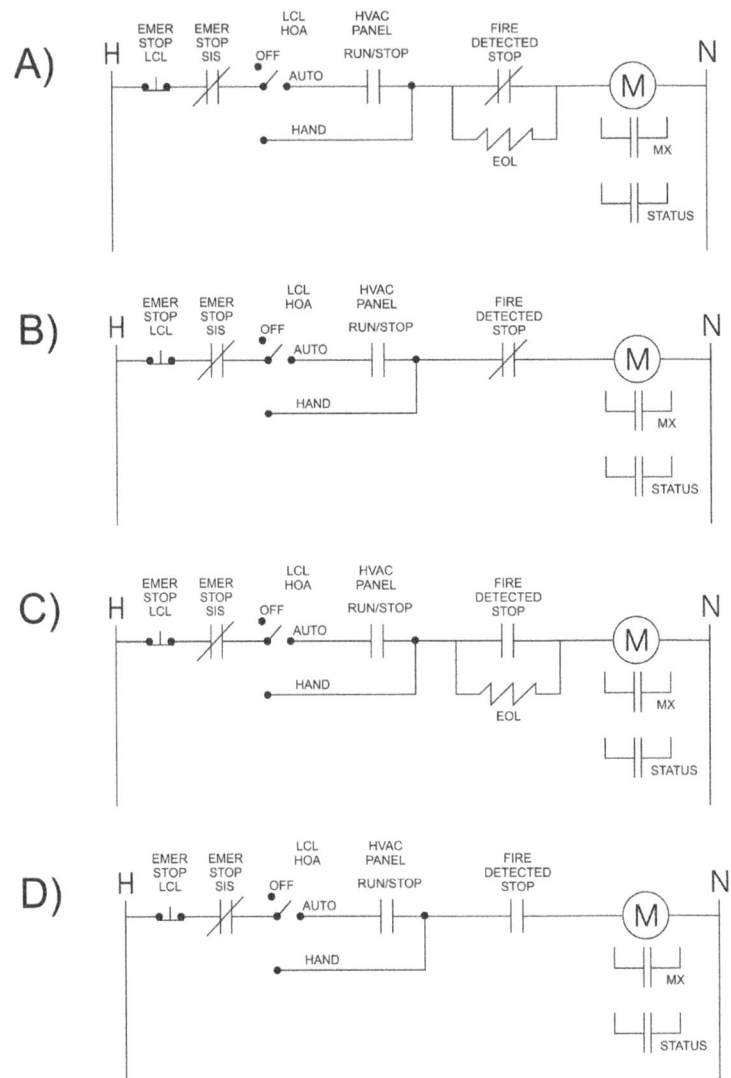

124. According to ANSI/ISA-5.1-2009, *Instrumentation Symbols and Identification*, which of the following is the symbol for a pressure-reducing regulator with an external pressure tap?

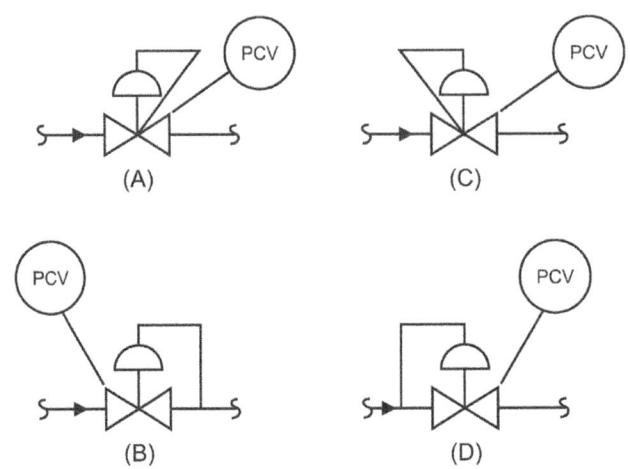

125. The logic diagram on the next page depicts the control scheme for an acid injection pump used to control the pH in a plant water system. The control of the pump is performed exclusively by a programmable logic controller (PLC). The hand-off-auto (HOA) local hand switch is spring return to OFF from the HAND position.

The combination of OR-gate "C" with AND-gate "D" is commonly referred to as a(n):

(A) Flip-flop circuit

(B) Exclusive-OR circuit

(C) Seal circuit

(D) Inverter

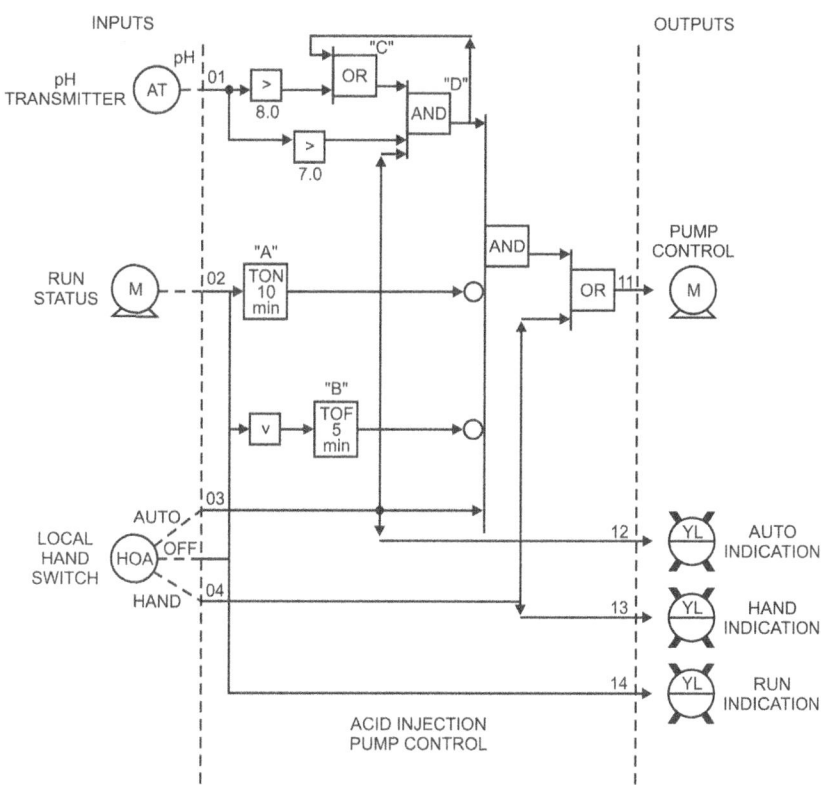

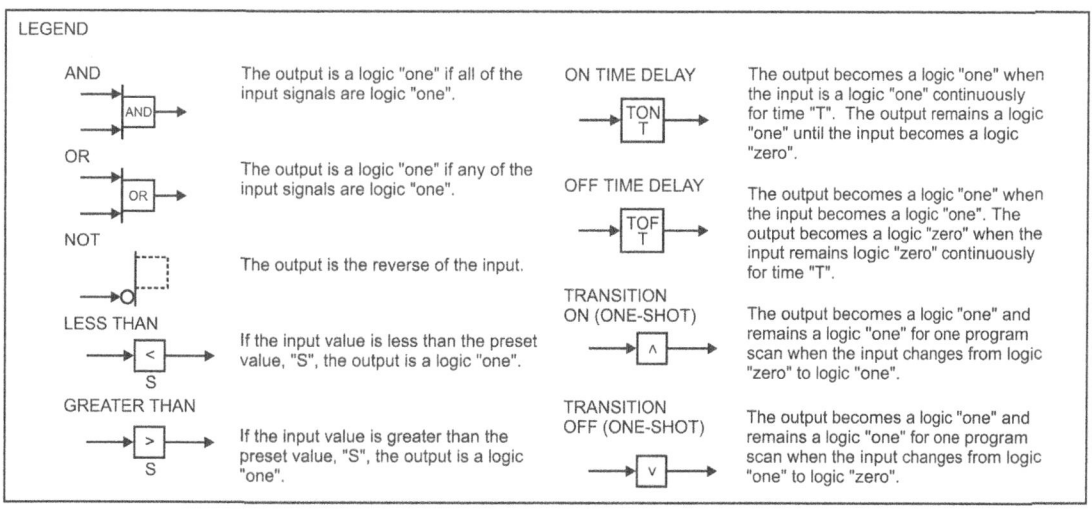

126. Refer to the sketch shown below, where two liquid feed streams are combined for a certain mixing operation to produce a single stream (product).

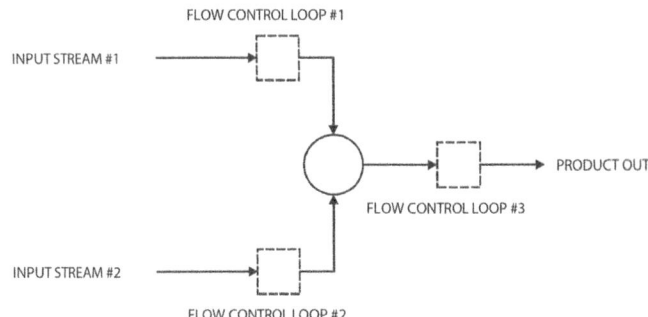

Based on the conservation requirements, what are the degrees of freedom of this system?

(A) One degree

(B) Two degrees

(C) Three degrees

(D) None of the above

127. The response of a system to a 1-unit step input is shown below.

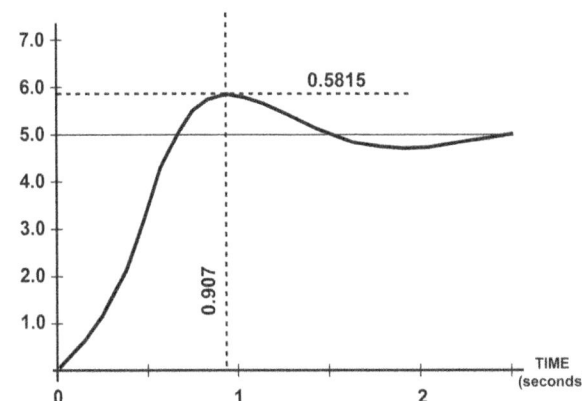

The data obtained from this test include:

Peak value of unit step response = 0.5815

Steady-state value of unit step response = 0.500

Time to peak value of unit step response = 0.907 s

Assuming that the system can be described by a second-order differential equation, the damping ratio is most nearly:

(A) 0.4   (B) 0.5

(C) 0.6   (D) 0.7

128. A variable-speed drive is used to control the rotary bin dispenser to a loss-in-weight feeder. Estimate the integral time setting and period of the resulting damped oscillation that minimizes the integrated absolute error (IAE) from a load disturbance for a process with the following dynamics:

1. Process dead time = 0.3 min

2. 20% change in the feeder set point causes a 28% change in mass flow (dW/dt)

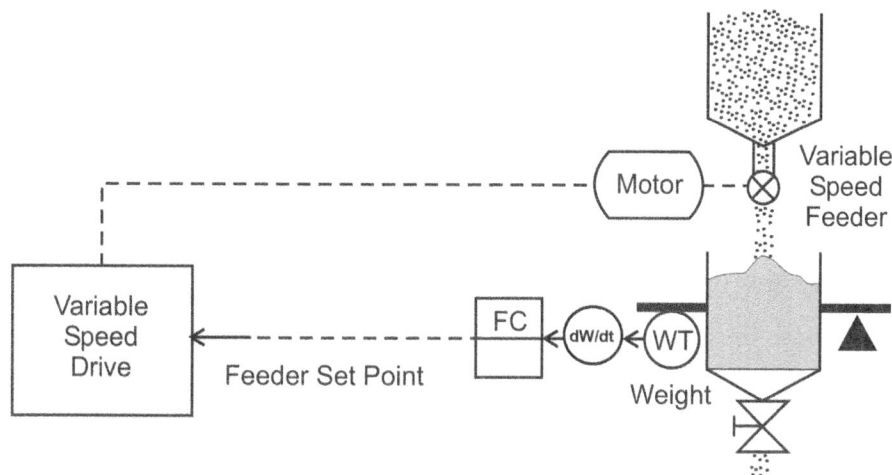

3. Process time constant = 0 min

4. Negligible measurement and feeder response times and controller execution times

(A) I = 45 s, τ_0 = 9 s

(B) I = 23 s, τ_0 = 129 s

(C) I = 1 s, τ_0 = 2 s

(D) I = 45 s, τ_0 = 129 s

129. According to ISA-5.5-1985, *Graphic Symbols for Process Displays*, concerning graphic symbols for use on visual display units (VDUs), the generic term for cathode-ray tube or solid-state display devices, which of the following statements is true?

I   Materials flows must be depicted in a left-to-right direction.

II   Symbols for unimportant process equipment can be omitted in displays.

III   An outline symbol indicates a stopped or inactive status, and a solid (filled) symbol indicates a running or active status.

IV   Red means *closed* or *off*, whereas green means *open* or *on*.

(A) I and III   (B) I and IV

(C) II and III   (D) I, II, III, and IV

130. Consider the application of a hydrogen sulfide (H_2S) analyzer that a plant is installing for personnel protection.

    The maximum permissible level of H_2S for an 8-hour (weighted average) exposure is 20 ppm (parts per million). A dose of 150 ppm or more for a short time can cause permanent injury or possibly death. The human nose can detect 1 ppm levels but is desensitized quickly with continuing exposure.

    The range of the H_2S analyzer is 0–20 ppm. The most appropriate setting for the alarm point (in ppm) would be:

    (A) 0–4.9   (B) 5.0–9.9

    (C) 10.0–14.9   (D) 15.0 or higher

131. A poorly tuned PID control loop has consistently responded too slowly to load disturbances because of a noise-free but slow measurement device. An appropriate action to accelerate the disturbance response would be to:

    I   Increase the controller's gain (decrease the proportional band).

    II   Increase the controller's integral action (increase the reset rate).

    III Decrease the controller's integral action (decrease the reset rate).

    IV   Increase the controller's derivative action (increase the derivative time).

    (A) I and II   (B) I and III

    (C) III   (D) IV

132. A poorly tuned PID control loop has consistently overcompensated for low-humidity conditions in a humidity-controlled room by injecting too much steam into the air supply duct. An appropriate action to reduce the overshoot problem would be to:

    (A) Retune the loop for a critically damped step response.

    (B) Retune the loop for an under-damped step response.

    (C) Retune the loop for a 1/4-wave decay step response.

    (D) Retune the loop to minimize rise time.

133. What is the best definition of a safety integrity level (SIL)?

    (A) A percentage (between 0 and 100) used to define the availability requirements of a safety instrumented function

    (B) A discrete number (1 through 4) used to define the performance requirements of a safety instrumented function

    (C) A probability (between 0 and 1) used to define the likelihood of a dangerous failure of a safety instrumented function

    (D) A percentage (between 0 and 100) used to define the ratio of the safe failure rate and the total failure rate

134. What is the *primary* reason for recording safety instrumented system testing and inspections?

(A) To prove that testing and inspections are actually being done

(B) To re-evaluate failure rate data as compared to original assumptions

(C) To verify that systems have not undergone unauthorized changes

(D) To satisfy the requirements of the process safety management regulation

135. What is the best definition of a safety life cycle?

(A) The total length of time that a device remains functional

(B) The activities involved in the development of the safety requirements specification

(C) The total length of time that a process facility remains operational

(D) The necessary activities involved in the implementation of safety instrumented functions

136. What are the best factors to base safety system test intervals on?

(A) Corporate policies, regulatory requirements, process operating parameters

(B) Performance targets, equipment failure rates, equipment configuration/redundancy

(C) Manufacturer recommendations, environmental factors, age of the plant

(D) Recommendations of peers, production requirements, bypass capabilities

137. Which of the following is the best reason for performing an MOC?

(A) Recording all changes made to systems for tracking purposes

(B) Ensuring that documentation reflects the actual plant design

(C) Informing upper-level management of changes made to systems

(D) Ensuring that safety is maintained despite changes made to systems

138. Which configuration offers the best safety performance (i.e., the *lowest* probability of failure on demand—PFD) for a safety instrumented function (SIF), *not* accounting for common cause?

(A) 1oo1   (B) 1oo2

(C) 2oo2   (D) 2oo3

139. An industrial control network has been shut down by a distributed denial-of-service attack (DDoS). Which security objective is affected?

    (A) Reliability

    (B) Integrity

    (C) Confidentiality

    (D) Availability

140. Coal grinding areas in a coal-fired steam power plant would be classified under the hazardous area provisions of the National Electric Code (NEC) as:

    (A) Class I, Group B

    (B) Class I, Group D

    (C) Class II, Group E

    (D) Class II, Group F

Second Session

501. A differential pressure transmitter (LT-100) is used to monitor the level in a horizontal storage vessel (V-100) that contains hot water. The vessel is manually filled and drained by valves V-1 and V-2. The pressure taps on the vessel for LT-100 are 60 in apart and the transmitter is mounted 10 in below the bottom process tap. The transmitter's process tubing is routed 10 ft horizontally before dropping vertically to the transmitter.

The following data apply:

Ambient temperature: 80°F (always)

Normal operating pressure: 100 psig (PIC-101)

Normal operating temperature: 280°F (TIC-102)

Seal fluid for the reference leg of the level transmitter:

SG   = 0.8 at 60°F

            = 0.62 at 80°F

            = 0.32 at 280°F

Boiling point = 520°F at 0 psig

The other leg of the level transmitter contains process fluid.

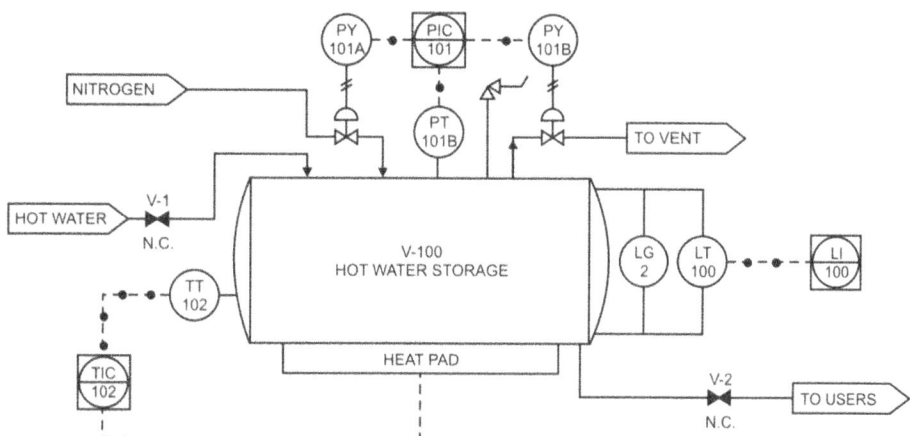

If LI-100 initially reads 50% and the pressure increases by 10%, the new reading of LI-100 at steady-state conditions would be:

(A) 0%   (B) 40%

(C) 50%   (D) 60%

502. To avoid flooding a distillation column, a differential pressure measurement is required across the top half of the column. The fluid entering the bottom of the column is similar to heavy crude oil. Due to the corrosive nature of the fluid, the bottom half of the column is clad with alloy steel. The bottom temperature is 1200°F. The product from the top of the column is light hydrocarbons at 250°F.

The first available measurement point is on a platform at an elevation of 25 ft above grade. The elevation of the platform at the top of the column is 100 ft. Access to all platforms is by means of ladders.

What transmitter(s) should be used?

(A) One differential pressure transmitter

(B) One differential pressure transmitter and two temperature transmitters

(C) Two temperature transmitters

(D) Two pressure transmitters

503. The flow of water in a 4-inch steel pipe is measured with an orifice plate and a differential pressure transmitter. At a flow rate of 120 gallons per minute (gpm), the differential pressure is 27 in of water. At a flow rate of 176 gpm, the differential pressure will be most nearly equal to:

(A) 12   (B) 18

(C) 39   (D) 58

504. For measuring the flow of raw sewage in a 4-inch steel pipe at a flow rate of 150 gpm, which of the following sensing devices will provide the most reliable and maintenance-free installation?

(A) Coriolis flowmeter

(B) Magnetic flowmeter

(C) Orifice plate

(D) Ultrasonic flowmeter

505. A tank level is measured using a differential pressure transmitter and a bubbler tube. The tank is vented to the atmosphere. The bubbler tube bottom is 1 ft above the tank bottom; the tank wall is 20 ft high. A 0 to 10 psi differential pressure gauge, accurate to 0.25 percent of full scale, is connected to the bubbler tube connection at the high-pressure side of the transmitter. The low-pressure side of the transmitter is connected to the tank top.

When the water level in the tank is 14 ft, the gauge reading in pounds per square inch (psi) is most nearly equal to:

(A) 5.6   (B) 6.1

(C) 6.5   (D) 13.0

506. Which of the following statements are true?

The accuracy of orifice-type flow elements is affected by:

    I   Upstream piping configuration

    II Downstream piping configuration

    III Eccentricity of the internal diameter of the meter run

    IV Entrained gas or air bubbles

    V Erosion of the hole bored in the orifice plate

(A) I and V only

(B) II and IV only

(C) I, II, and V only

(D) I, II, III, IV, and V

507. For a gas-fired heater, the best choice for flame detection is a(n):

(A) Flame rod detector

(B) Infrared detector

(C) Silicon cell detector

(D) Self-check UV detector

508. Consider the application of a hydrogen sulfide (H_2S) analyzer that a plant is installing for personnel protection.

Assuming no mechanical problems or other limitations at the site, where should a field sensor be placed to provide the earliest possible warning of an excessive concentration of H_2S?

(A) 1 ft above ground

(B) 3 ft above ground

(C) Eye level

(D) 15 ft above ground

509. In an area of high electromagnetic disturbances, the computer-data transmission medium with the least noise pickup is:

(A) Fiber optics

(B) Twisted pairs of copper conductors with shielding

(C) Twisted pairs of copper conductors in conduit

(D) Coaxial cable

510. In hazardous areas, intrinsically safe circuits can be wired:

    (A) Only in metal conduit

    (B) Only in sealed or vented cables

    (C) Only with automatic shutdown on electric power failure

    (D) In the same way as in nonhazardous areas

511. Comparing unshielded twisted pairs (UTPs) with shielded twisted pairs (STPs) for data transmission in a control system, which of the following statements is false?

    (A) UTP is less expensive for materials and installation.

    (B) STP is more resistant to electrical interference.

    (C) UTP and STP are equally acceptable to vendors of LANs.

    (D) UTP and STP can both be used at 5 million bits/s.

512. Control valve noise reduction can be achieved through source treatment or through path treatment. Which of the following is *not* a method of path treatment?

    (A) Extra acoustic insulation

    (B) Barricades and warning signage

    (C) Heavy wall pipe

    (D) Diffusers/silencers/mufflers

513. What is *not* a function of a network firewall in an industrial network application?

    (A) Limiting bandwidth usage

    (B) Limiting unauthorized data access

    (C) Separation of network layers

    (D) Separation of control and safety systems

514. Based on the following bathtub curve, when should safety instrumented function (SIF) devices be removed from service and replaced?

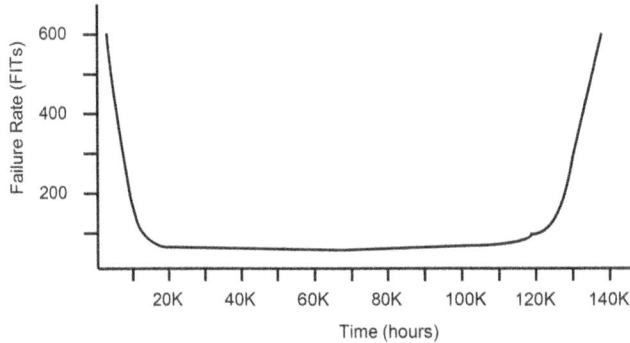

    (A) At 120,000 h   (B) At 10,000 h

    (C) At 100 FITs   (D) At 200 FITs

515. A control valve, originally supplied for gaseous service, is now being considered for a liquid service application.

Original service conditions:

Maximum capacity = 60,000 scfh

ΔP for valve sizing = 50 psi

Gas molecular weight = 44

Inlet pressure = 300 psig

Inlet temperature = 120°F

New service conditions:

ΔP for valve sizing = 10 psi

Liquid specific gravity = 0.81

Inlet pressure = 240 psig

In the new service, assuming the maximum C_v = 12, the maximum flow in gallons per minute (gpm) is:

(A) 30.7   (B) 34.2

(C) 37.9   (D) 42.2

516. Consider a gas flow control loop in manual, with the initial process conditions (A = upstream and B = downstream) as given in the figure below. All conditions remain constant other than the changes specified in each question. Subscripts 1 and 2 refer to the old and new conditions, respectively.

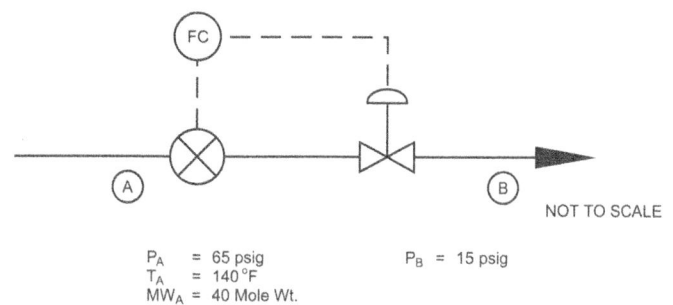

P_A = 65 psig P_B = 15 psig
T_A = 140 °F
MW_A = 40 Mole Wt.

If only the open flow area (X) of the valve increased, which of the following best describes how the mass flow (F) would change?

(A) $F_2 = F_1(X_1 / X_2)^{0.5}$

(B) $F_2 = F_1(X_2 / X_1)^{0.5}$

(C) $F_2 = F_1(X_2 / X_1)$

(D) $F_2 = F_1(X_2 / X_1)^2$

517. Which of the following statements about control valve installation practices is false?

(A) In locating valves, consideration must be given to the manufacturer's recommended clearances and also to the positions of heating ducts and electrical wire ways.

(B) For safety reasons, valves must always go to a closed or open position on power failure.

(C) To allow for plant expansion, it is common to initially select valves smaller than the connecting pipe.

(D) Characteristics of the flowing material determine whether flanged, threaded, or welded pipe joints can be used.

518. You are to engineer a motor control system in a relay-based motor controller in the DCS with the following requirements:

1. Run status is to be displayed on the operator HMI.

2. The motor is NOT to automatically restart after a power failure.

3. You have one DI and two DO available.

How are the I/O to be configured and wired into the motor control circuit?

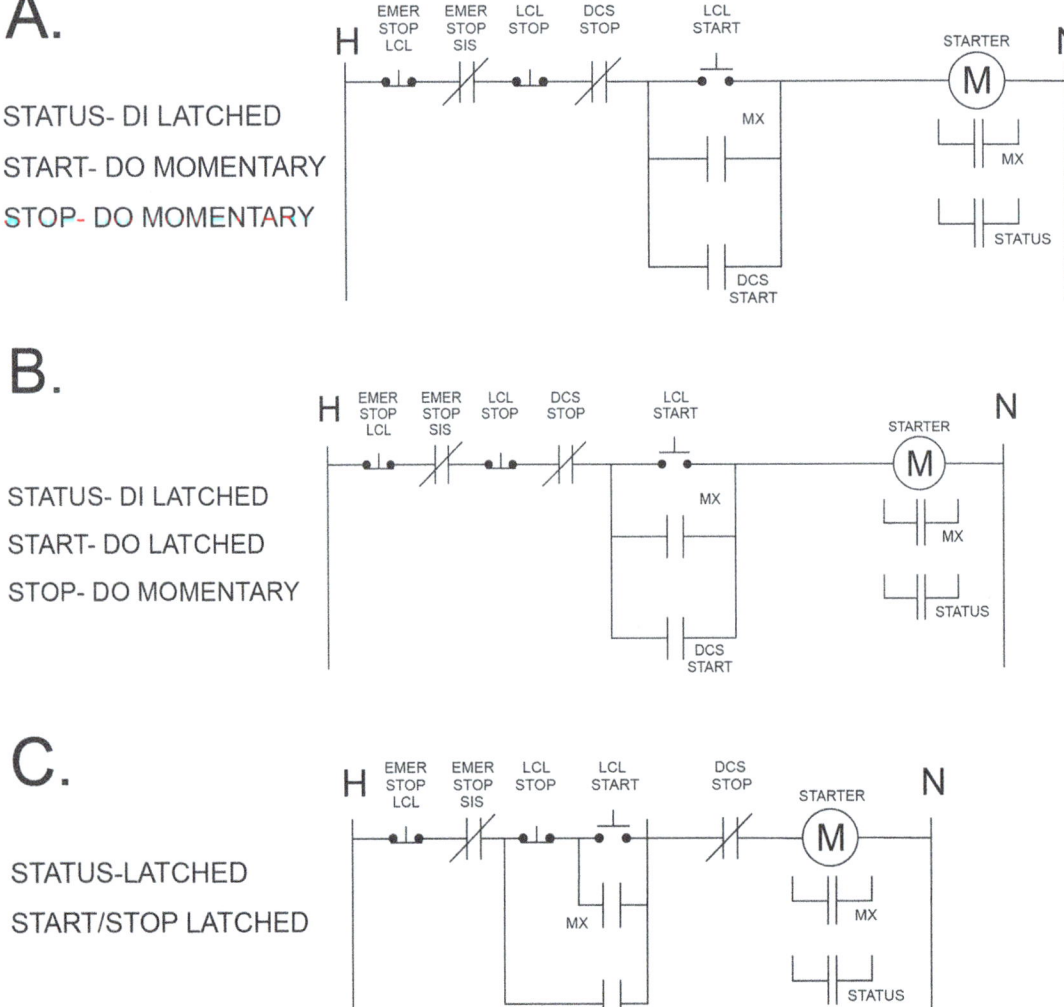

D.

STATUS-LATCHED

START/STOP LATCHED

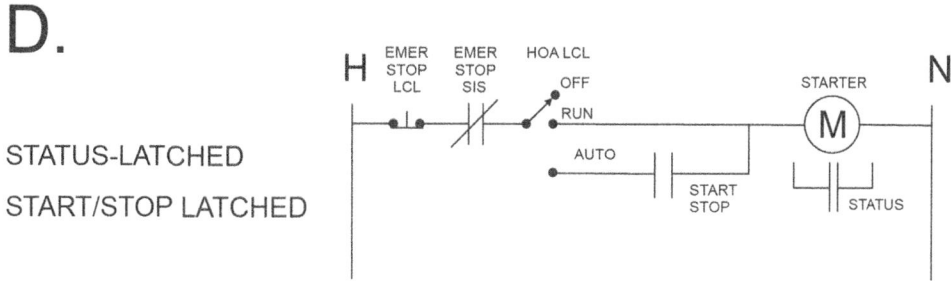

519. A new pump was installed with the motor control, shown in the figure below, with a low flow shutdown. The wiring and pump rotation were confirmed during checkout. When operations went to place the pump in service, the pump would not start. What is the most likely cause?

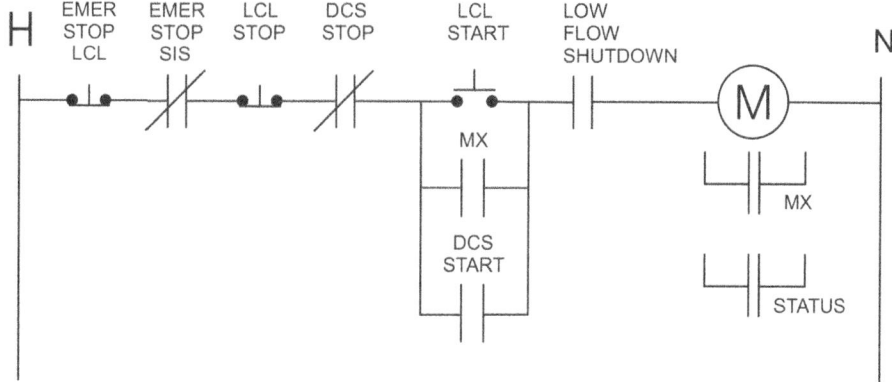

(A) The low-flow switch must have a normally closed start-up bypass with a time-delay-on relay.

(B) The flow switch must have a normally closed start-up bypass with a time-delay-off relay.

(C) The low-flow switch must have a normally open start-up bypass with a time-delay-off relay.

(D) The low-flow switch must have a normally open start-up bypass with a time-delay-on relay.

520. Double-acting control valve actuators usually:

(A) Fail in the closed position

(B) Fail in the open position

(C) Fail in the last position

(D) Are not fail-safe

521. Which of the following diagrams shows the best scheme for double-acting, fail-last-position actuation of the process valve?

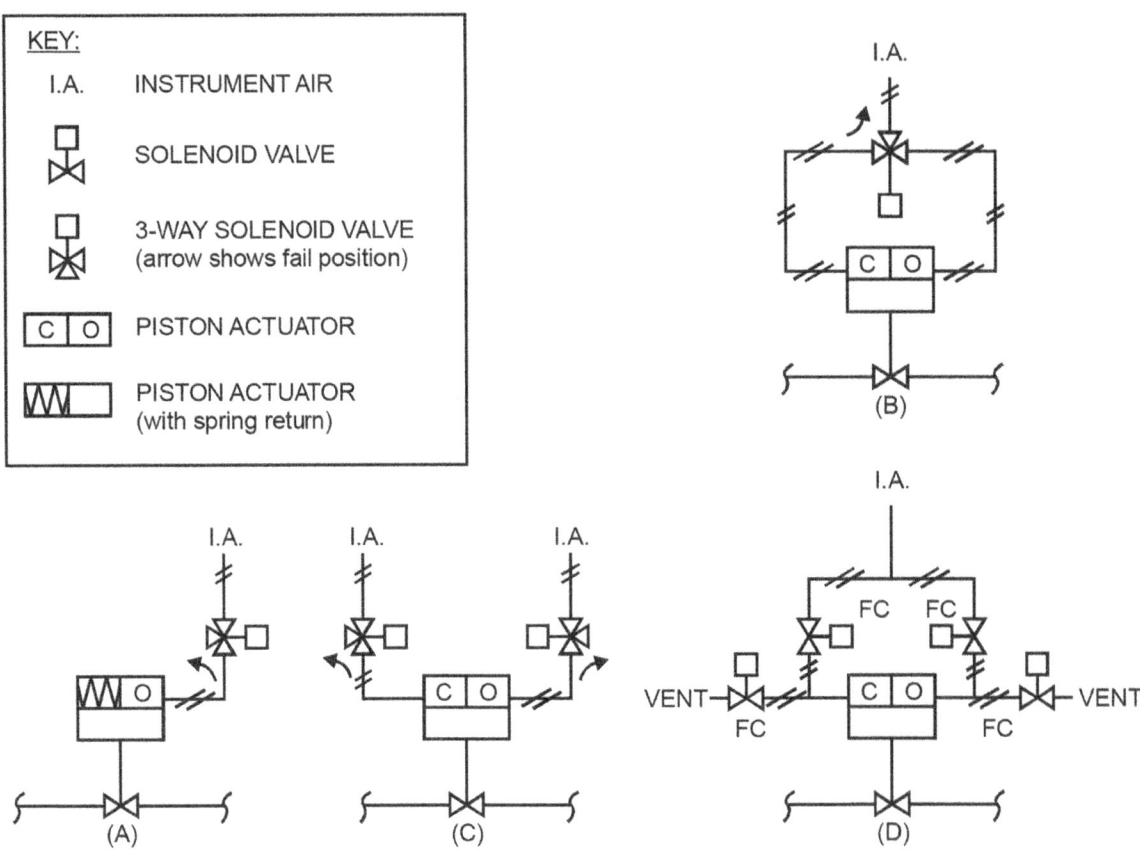

522. According to ANSI/ISA-5.1-2009, *Instrumentation Symbols and Identification*, which of the following is the symbol for a discrete instrument, not accessible to an operator, in an auxiliary location?

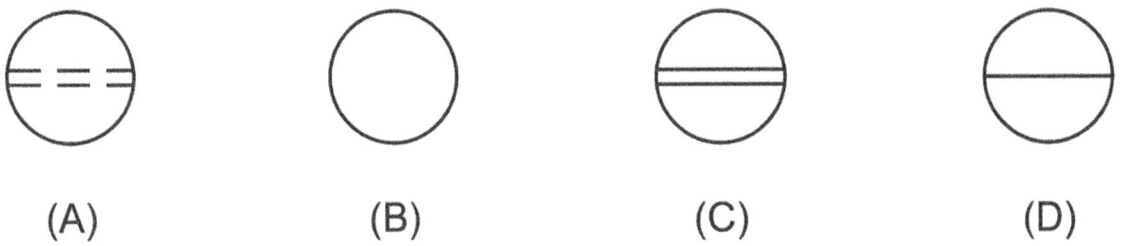

(A)   (B)   (C)   (D)

523. Which of the following Boolean statements (the notation " + " means *OR* and "•" means *AND*) describes the operation of the logic circuit shown in the following diagram?

    (A) M = A + (B • C • D)

    (B) M = A + B + (C • D)

    (C) M = A • (B + (C • D))

    (D) M = A + (B • (C + D))

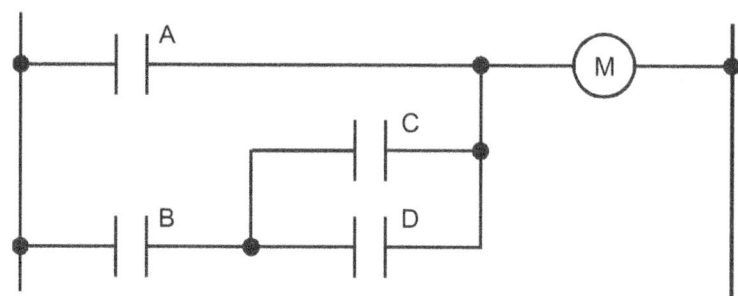

524. A control system is described by the block diagram shown below.

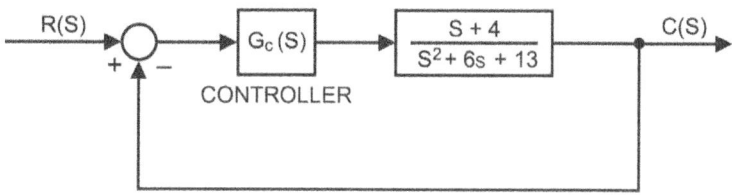

CONTROLLER

NOT TO SCALE

Data list:

$G_s(s)$ = system transfer function = $\dfrac{s+4}{s^2+6s+13}$

$G_c(s)$ = controller transfer function = $\dfrac{K(s+3)}{s(s+1)}$

For which values of K is the system stable?

(A) All $K > 0$ (B) All $K < 0$

(C) All $K > 19$ (D) All $K < 13/7$

525. An alternative to open-loop process testing for determining controller parameters is closed-loop testing: the controller is placed in automatic, the integral and derivative actions are set to zero, and the controller gain is increased gradually until a sustained oscillation is produced. Which of the following statements about this procedure are correct?

    I   The two data items obtained from a closed-loop test and used in tuning parameter calculation are the period and amplitude of oscillation.

    II   The tuning parameters obtained from a closed-loop test are likely to be more accurate than those determined from an open-loop test.

    III It is necessary to observe the process for many cycles (say, 10 or more) to be sure that the oscillation is neither decaying nor increasing.

    IV The engineer conducting a closed-loop test has no control over the amplitude of the oscillation.

(A) I and III   (B) I, II, and IV

(C) II and III   (D) II and IV

526. Compared to a control loop with no dead time (pure time delay), a control loop with an appreciable dead time tends to require:

(A) Less proportional gain and less integral action

(B) More proportional gain and less integral action

(C) More proportional gain and more integral action

(D) Less proportional gain and more integral action

527. Using the ISO 5024 standard conditions of $P_{base} = 14.69595$, $T_{base} = 59°F$ for natural gas, a mass flow of 21,755 lb/h of gas at 60 psig and 90°F to standard cubic feet per hour (scfh) is most nearly:

Molecular Weight = 28.962

Compressibility = 0.999

(A) 32,349 scfh   (B) 220,318 scfh

(C) 284,188 scfh   (D) 301,378 scfh

528. Which of the following tuning criteria would be most appropriate for designing a controller to regulate the temperature in the room where you are now sitting?

(A) Minimize the response time for set-point changes.

(B) Minimize the overshoot for set-point changes.

(C) Follow varying set points with a minimum error.

(D) Maintain the controlled variable at a constant value.

529. A differential pressure transmitter is used to measure and control the oil/water interface level in the boot of a coalescer vessel. The level measurement taps on the boot are 14 in apart. There is no rag layer present. The transmitter is close-coupled to the lower tap and uses a remote diaphragm seal (seal fluid SG = 0.934) for the upper tap.

Oil SG = 0.82

Water SG = 0.99

What should be the calibrated range of the transmitter?

(A) 0.784 to 1.596 in wc (B) − 1.596 to 0.784 in wc

(C) 13.2 to 14.84 in wc   (D) 11.36 to 13.2 in wc

530. What effect does timer "A" have on the pump control?

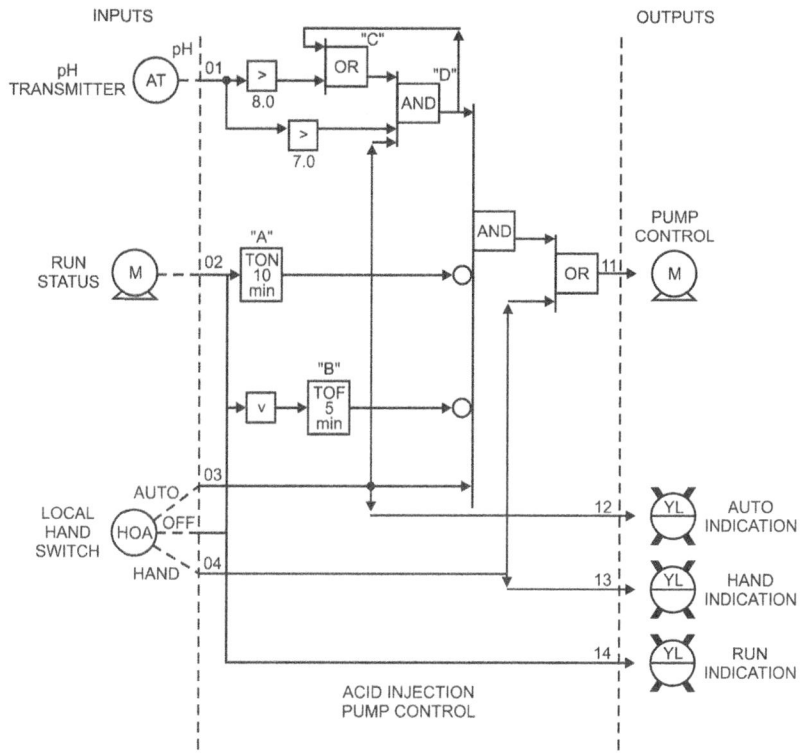

ACID INJECTION
PUMP CONTROL

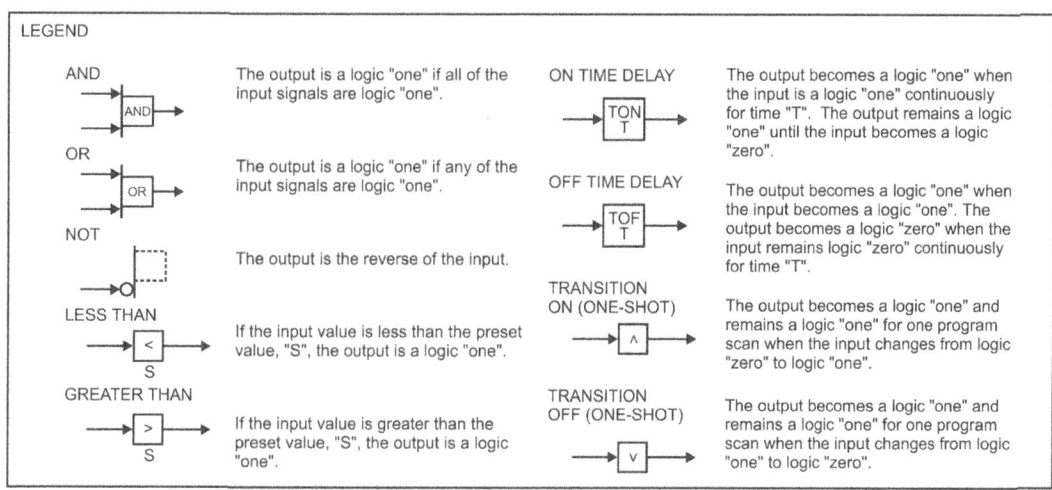

(A) It starts the pump 10 min after the pH reaches 7.0.

(B) It stops the pump after it has run for 10 min.

(C) It automatically starts the pump every 10 min.

(D) It delays the pump start for 10 min.

531. Which ISA standard provides requirements for the specification, design, installation, operation, and maintenance of safety instrumented systems?

(A) 5.1   (B) 50

(C) 84   (D) 88

532. What is the objective of factory acceptance testing of a safety instrumented system?

(A) To test that the hardware and software satisfy the requirements defined in the safety requirements specification

(B) To test and document that the facility fabricating the system is qualified to perform the required work

(C) To test that field devices are operational before placing them in service

(D) To create test procedures that will be used during the life of the system to verify functionality

533. The fundamental goal of any safety integrity level (SIL) selection study is to ensure that:

(A) All risks are designed out of the process.

(B) Equipment is operated at peak efficiency.

(C) The residual risk is at or below the tolerable risk.

(D) The tolerable risk is at or below the residual risk.

534. A control system employs three sensors, each having a failure probability of 0.02 in 6 months of operation. The system can function properly when any two or more of the sensors are working, but it must shut down if two or three of the sensors fail.

The probability that the system can operate for 6 months without a shutdown is most nearly equal to:

(A) 0.9412   (B) 0.9600

(C) 0.9800   (D) 0.9988

535. Pressure-temperature compensation is often used to convert flowing volumetric flow rate to volumetric flow rate at standard conditions. Which of the following equations correctly represents PT compensation for a gas (where b represents base conditions and f represents flowing conditions)?

(A) $Vb = Vf \cdot \left(\dfrac{Pb}{Pf}\right) \cdot \left(\dfrac{Tf}{Tb}\right) \cdot \left(\dfrac{Zb}{Zf}\right)$

(B) $Vb = Vf \cdot \left(\dfrac{Pb}{Pf}\right) \cdot \left(\dfrac{Tb}{Tf}\right) \cdot \left(\dfrac{Zf}{Zb}\right)$

(C) $Vb = Vf \cdot \left(\dfrac{Pf}{Pb}\right) \cdot \left(\dfrac{Tb}{Tf}\right) \cdot \left(\dfrac{Zb}{Zf}\right)$

(D) $Vb = Vf \cdot \left(\dfrac{Pf}{Pb}\right) \cdot \left(\dfrac{Tf}{Tb}\right) \cdot \left(\dfrac{Zf}{Zb}\right)$

536. The maximum allowed length of a nonintrinsically-safe FOUNDATION Fieldbus segment (trunk + spurs) provided the cable meets the specification requirements is:

(A) 1600 m   (B) 1900 m

(C) 2000 m   (D) 2600 m

537. MOC is not required in which case?

(A) Low-risk applications

(B) Replacements in kind

(C) Changes to software

(D) Changes in personnel

538. Which of the following statements about the installation of a tank blanketing pressure regulator is false?

(A) Piping from the regulator to the headspace of the tank should be minimized.

(B) The regulator should be set to control at the tank maximum allowable working pressure (MAWP).

(C) The sensing tap for the regulator should be on a separate section of the tank roof from the regulator outlet piping connection.

(D) Failure of the regulator contingency is considered when sizing tank vents and emergency relief vents.

539. Your assignment is to implement a burner management system using a solid-state, microprocessor-based programmable logic controller (PLC). You will analyze boiler operation, develop system logic, and select control equipment.

The system design shall upgrade the burner management functions on a 50,000,000 Btu/h, two-burner, gas-fired boiler to current NFPA standards. Igniters will be interrupted (turned off) when the timed trial for ignition of the main burner expires. PLC inputs and outputs will de-energize to shut off (trip) fuel to the boiler.

The burner-system manual emergency shutdown shall be accomplished by:

(A) A normally open contact to a PLC input.

(B) A normally closed contact to a PLC input.

(C) A normally open contact to a PLC input and a normally closed contact to another PLC input.

(D) A hard-wired, normally closed contact to de-energize a fuel trip relay.

540. Which of the following statements regarding fiber-optic cable usage in chemical plants is false?

(A) Fiber-optic cables are not a potential ignition source in hazardous areas.

(B) Single-mode fiber can provide approximately 20 times the distance over multimode fiber at the same data rates.

(C) Short fiber-optic cable runs may require the use of optical attenuators to avoid receiver saturation.

(D) Optical time-domain reflectometry (OTDR) testing can detect faults and characterize individual fibers.

Appendix C
Answers to Sample Questions

This appendix contains solutions and/or answers to the questions contained in Appendix B.

A one-page summary list of the answers is included to facilitate scoring for those who treat the questions in Appendix B as a sample exam.

Comments about the questions and answers are appreciated. Please send them to the ISA contact below and they will be forwarded to the CSE exam committee.

Director of Credentialing Services, ISA
P.O. Box 12277
Research Triangle Park, NC 27709

Answers

Question	Answer	Question	Answer	Question	Answer	Question	Answer
101	D	121	B	501	C	521	D
102	C	122	B	502	D	522	A
103	C	123	A	503	D	523	D
104	C	124	B	504	B	524	A
105	C	125	C	505	A	525	D
106	A	126	B	506	D	526	A
107	C	127	B	507	D	527	C
108	C	128	D	508	A	528	D
109	C	129	C	509	A	529	B
110	D	130	B	510	D	530	B
111	A	131	D	511	C	531	C
112	B	132	A	512	B	532	A
113	D	133	B	513	D	533	C
114	C	134	B	514	A	534	D
115	D	135	D	515	D	535	C
116	C	136	B	516	C	536	B
117	A	137	D	517	B	537	B
118	A	138	B	518	A	538	B
119	A	139	D	519	D	539	D
120	C	140	D	520	D	540	A

Answers: First Session

101. The correct answer is (D).

For this control application, the best choice is an orifice plate.

102. The correct answer is (C).

The "90% point" is a boiling characteristic of a hydrocarbon liquid. Thus, the only suitable analyzer is a boiling point analyzer. The other instruments measure different characteristics of materials.

103. The correct answer is (C).

$$\text{Span} = SG \cdot d$$

$$SG = \frac{Density\ of\ Process\ at\ Normal\ Temp = 280°F}{Water\ Density\ at\ Standard\ Temp = 60°F}$$

$$SG = \frac{57.941\ \#/ft^3}{62.364\ \#/ft^3} = 0.929$$

$$\text{Span} = 0.929 \cdot 60\ \text{in} = 55.7\ \text{in}\ H_2O$$

104. The correct answer is (C).

105. The correct answer is (C).

The 100 ft of head produced by the pump would be 43.3 psig of water (SG = 1.0), but the pumped material is 0.8 SG, so the answer is: 100 ft water (0.433 psig/ft water) • 0.8 = 34.6 psig. The result must be added to the 100 psig at the suction to yield 134.6 or 135 psig.

106. The correct answer is (A).

There are 42 gallons in a US barrel, which will be used for the conversion: at 42 gal/barrel, 30,000 barrels = 1,260,000 gal to be transferred. The transfer is to be made in 6 h or 360 min; therefore, the flow rate must be 1,260,000 gal/360 min, or 3500 gpm.

107. The correct answer is (C).

108. The correct answer is (C).

109. The correct answer is (C).

Zener diode barriers are inexpensive and easy to use. The other statements are true.

110. The correct answer is (D).

The 4–20 mA signal will require a 250-ohm resistor to provide a 1–5 V input to the indicator. To ensure no adverse effects to the control signal, based on the input impedance, an I:I repeater should be used.

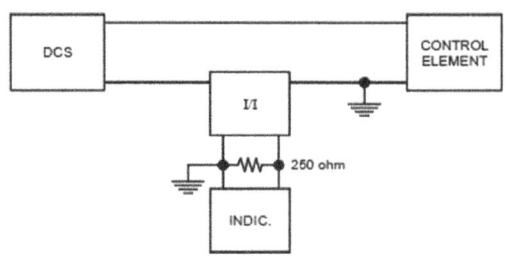

111. The correct answer is (A).

Wireless systems with many links or channels in a limited radio frequency spectrum can interfere with each other; there is no limit on the possible number of wired or cable channels.

(B) is true. Wireless systems do not require conduit or cable trays to be routed through the plant in order to connect a multitude of inputs to controllers and central processors; many functions are built into the wireless system components.

(C) is true. Sensors and controllers in a wireless system still need power to perform their functions.

(D) is true. By means of channel hopping, encryption, resending data packets, and other methods, wireless systems can be given a high degree of immunity to electrical noise.

112. The correct answer is (B) because the fiber can be broken by a small-radius bend.

113. The correct answer is (D).

A fieldbus installation will have less total wiring due to the combining of multiple signals on a single cable instead of having all the transmitters wired individually into a central control room.

114. The correct answer is (C).

The ring topology will not tolerate a fault due to its structure. Therefore, answers A, B, and D are incorrect. Selection C is the correct answer. Star and mesh configuration can handle faults.

115. The correct answer is (D).

$$C_v = \frac{Q_G}{1360 P_1 Y} \sqrt{\frac{GTZ}{X}}$$

where

Q_G = gas flow in standard cubic feet per hour (scfh) = 60,000

T = gas temperature in degrees

Rankine = 580 degrees R

(degrees Fahrenheit plus 460)

G = specific gravity (SG) of the gas = 1.52

Z = compressibility factor (1.0 for pressures less than 100 psia)

Y = expansion factor

Y = $Y = 1 - \dfrac{X}{3X_T} = 1 - 0.16/3\,(0.61) = 0.912$

X = pressure drop ratio $X = \dfrac{\Delta P}{P_1} = 50/315 = 0.16$

X_T = $0.85F_L^2$ (F_L depends on valve style) = 0.85 $(0.85)^2$ = 0.61

(0.85 for globe valves and 0.6 for ball valves)

P_1 = inlet pressure (psia) = 315 psia

The gas sizing equation may be simplified when a certain condition exists: When X is less than 0.1, and in this case the X value is only 0.16, a simplified solution may be obtained.

$$C_v = \frac{Q_G}{963}\sqrt{\frac{GT}{\Delta P(P_1 + P_2)}} = (60{,}000/963) \times sq.rt.\left[(1.52 \times 580)/50 \times (315 + 265)\right]$$

C_v = 10.8 and the closest value is 10.4.

where

ΔP = $P_1 - P_2$ = 50 psi

P_2 = outlet pressure (psia) = 265 psia

G = SG (If the molecular weight of the gas is known, the SG may be calculated by dividing by the molecular weight of air by 29.)

SG = 44/29 = 1.52

116. The correct answer is (C).

Putting valve stems in a vertical position is a common practice but is not an absolute requirement for all situations. (See J. W. Hutchison, *ISA Handbook of Control Valves*, Second Edition, 1976, 353.)

117. The correct answer is (A).

Vent lines should be provided so that personnel or equipment will not be inadvertently sprayed, even with harmless material, when a vent valve is opened. (See J. W. Hutchison, 336–337.)

118. The correct answer is (A).

Ball valves usually have larger C_v for the same size body than globe valves.

(B) Piping might need to be revised, but ball valves are usually shorter than the same size globe valve.

(C) Ball valves often use piston actuators but can use diaphragm-type actuators as well.

(D) Ball valves are often used for on/off control but are also used for modulating control.

119. The correct answer is (A).

According to ASME Section VIII, Division 1, UG-134(d), *Pressure Setting of Pressure Relief Devices*: "The pressure at which any device is set to operate shall include the effects of static head and constant backpressure."

120. The correct answer is (C).

The maximum vessel pressure with one valve in service is 110% of MAWP (10% accumulation), or $1.1 \cdot 200 = 220$ psig.

121. The correct answer is (B).

The best approach is to keep the corrosive gas isolated from the relief valve by installing a higher metallurgy rupture disk so that a more economical material may be used for the valve. A reverse-buckling rupture disk is the preferred design as it provides many advantages over a standard rupture disk, including the ability to test the valve in service and to withstand a potential vacuum in the protected vessel. This would be the best and most accepted solution.

122. The correct answer is (B).

A single maintained contact is to be placed between the Emergency STOP SIS and the local STOP and after the local START.

123. The correct answer is (A).

The typical fire detection/protection panel outputs are normally closed and supervised. See NFPA-72, *National Fire Alarm and Signaling Code*. Therefore, motor control (A) is correct.

124. The correct answer is (B).

(A) is a self-contained, pressure-reducing regulator.

(C) is a self-contained, back-pressure regulator.

(D) is a back-pressure regulator with an external tap.

125. The correct answer is (C).

This arrangement and combination of an OR gate with the AND gate provides an output that serves as a seal-in input signal, after the initial trigger.

126. The correct answer is (B), two degrees of freedom. Solution:

Number of variables: 3   Number of equations: 1*

*Material balance: F3 = F1 + F2. Therefore, degrees of freedom = 3 – 1 = 2.

Any two of the flows can be specified; the third flow must satisfy the material balance.

127. The correct answer is (B).

The overshoot is

$$\frac{0.5815 - 0.500}{0.500} \times 100 = 16.3\%$$

From standard response curves, the damping ratio is approximately 0.5, one-half of critical damping. Alternately, the damping ratio (ξ) can be calculated from this equation:

$$16.3 = 100e^{-\pi\xi}/\sqrt{1-\xi^2}$$

$$e^{\pi\xi}/\sqrt{1-\xi^2} = \frac{100}{16.3} = 6.135$$

$$\pi\xi/\sqrt{1-\xi^2} = 1n6.135 = 1.814$$

$$\pi^2\xi^2 = 3.29(1-\xi^2); \xi^2(9.87+3.29) = 3.29;$$

$$\xi^2 = 0.25; \xi = 0.5$$

128. The correct answer is (D).

The equation for the integral time setting for integral-only control is in the book *Feedback Controllers for the Process Industries* (F. G. Shinskey) on page 158, CASE I, "Integral Alone with T1 = 0 (primary time constant is zero)."

$$\tau_d = 0.3 \text{ min } (60 \text{ s/min}) = 20 \text{ s}$$

$$K_p = 28/20 = 1.4$$

For integral-only control: $I = 1.6 \, K_p \tau_d = 45$

For $\tau_0 = 6.4 \, \tau_d = 129$

129. The correct answer is (C).

Statements II and III are true. Statement I is false because flows can be in any direction; recycle flows will be right-to-left if the main flows are left-to-right, and some flows are vertical, up or down. Statement IV is a common but not universal convention; the convention can vary from company to company or industry to industry.

130. The correct answer is (B).

The alarm point should be set above the noise or background level, perhaps caused by small, intermittent releases of H_2S, and well below the prescribed maximum permissible level. Of the choices given, the best is 5.0–9.9 ppm.

131. The correct answer is (D).

Derivative action in a relatively noise-free system causes the controller to anticipate the effects of a load disturbance and take preemptive action. Action I might also work, but there is nothing to indicate that action III is required as well.

132. The correct answer is (A).

All the other tuning objectives would require overshoot, which is stated to be undesirable. This question deals with important tuning criteria other than the classical 1/4-wave decay criteria favored by Ziegler and Nichols.

133. The correct answer is (B), based on clause 3.2.74 of ANSI/ISA-61511-1-2018/IEC 61511-1:2016 + AMD1:2017 CSV.

134. The correct answer is (B), to re-evaluate failure rate data. Failure rates and test intervals were assumed in calculations associated with the original design. These numbers may in fact not match both failure rates and test intervals in actual practice. These numbers must be compared to assure the original design is valid.

135. The correct answer is (D), based on clause 3.2.76 of ANSI/ISA-61511-1-2018/IEC 61511-1:2016 + AMD1:2017 CSV.

136. The correct answer is (B), based on clauses 16.3.1.3 and 11.9.2 of ANSI/ISA-61511-1-2018/IEC 61511-1:2016 + AMD1:2017 CSV.

137. The correct answer is (D), based on clause 17.1.1 of ANSI/ISA-61511-1-2018/IEC 61511-1:2016 + AMD1:2017 CSV.

138. The correct answer is (B).

    1oo2 has the lowest PFD. Two simultaneous failures would be required. With 2oo3, two simultaneous failures will result in a system failure, but there are three times as many dual failure combinations.

139. The correct answer is (D).

    Availability since service is denied. Reliability is not a security objective according to ANSI/ISA 62443-4-1-2018 and IEC 62443-4-1:2018.

140. The correct answer is (D).

    Class II Group F is for areas where coal dust is present.

Answers: Second Session

501. The correct answer is (C).

    Because the differential pressure measurement provides for pressure fluctuation and water is an incompressible fluid, the level measurement remains unchanged.

502. The correct answer is (D).

    If a single differential pressure transmitter was used, mounted at one of the platforms, the connecting lines would be significantly different in length. Extraneous effects (e.g., temperature or gravity head) could adversely affect the accuracy. Therefore, two pressure transmitters should be used.

503. The correct answer is (D).

$$\frac{\Delta P_2}{\Delta P_1} = \left(\frac{F_2}{F_1}\right)^2$$

$$\Delta P_2 = (27)\left(\frac{176}{120}\right)^2 = 58.1 \text{ in of } H_2O$$

504. The correct answer is (B).

For a stream containing solids, the best choice is a flow sensor that offers a minimum obstruction to the flowing stream, that is, a magnetic flowmeter.

505. The correct answer is (A).

The pressure reading is $(14-1)\left(\dfrac{62.4}{144}\right) = 5.6$ psi.

506. The correct answer is (D).

All the listed factors influence the accuracy of measurements made with orifice-type elements.

507. The correct answer is (D), self-check UV detector.

508. The correct answer is (A).

Hydrogen sulfide (MW = 44) is heavier than air (MW = 29); therefore, its concentration is highest close to the ground. For early detection, the field sensor (analyzer inlet) should be as close to the ground as possible, that is, 1 ft.

509. The correct answer is (A).

Fiber-optic data transmissions are immune to electromagnetic disturbances. The other media listed are electrical and therefore subject in varying degrees to electromagnetic disturbances.

510. The correct answer is (D).

That is the purpose of intrinsically safe equipment. Caution is needed at the boundary between hazardous and nonhazardous areas, however, to prevent leakage of, for example, explosive gases from the hazardous area to the nonhazardous area.

511. Answer (C) is false and is therefore the correct answer. Some LAN vendors insist on STP.

512. The correct answer is (B).

Path treatment methods include heavy wall pipe, extra acoustical insulation, and diffusers/silencers/mufflers. Barricades and signage do nothing to treat the noise.

513. The correct answer is (D).

514. The correct answer is (A), 120,000 h.

This is the time the wear failure rate increases due to wear-out failures.

(B) 10,000 h is the period in which burn-in failures have been eliminated.

(C) 100 FITs (failures in time, per billion hours) is the constant failure rate over the normal working life of the device. This is a failure rate, not a time interval of when the devices should be replaced.

(D) 200 FITs is a failure rate indicating more devices are failing compared to the constant failure rate during the normal life, but it does not indicate when devices should be replaced.

515. The correct answer is (D).

For the new service conditions, using the maximum $C_v = 12$, the maximum flow can be calculated using this equation (Equation 1, ANSI/ISA-75.01.01-2012 (60534-2-1 MOD) *Industrial-Process Control Valves – Part 2-1: Flow Capacity – Sizing Equations for Fluid Flow under Installed Conditions*):

$$C_v = Q\sqrt{\frac{G}{\Delta P}}$$

where

    C_v = valve flow coefficient

    G = fluid specific gravity

    Q = flow rate in gallons per minute (gpm)

    ΔP = pressure drop in pounds per square inch (psi) (inlet pressure minus outlet pressure)

$$q = C_v\sqrt{\frac{\Delta P}{G}} = 12\sqrt{\frac{10}{0.81}} = 42.2 \text{ gpm}$$

516. The correct answer is (C).

The mass flow is directly proportional to the open area of the valve, the density, and the fluid velocity.

    M1 = (ρ1 • A1 • V1): Mass flow is equal to the density times the area times the velocity.

    M2 = (ρ2 • A2 • V2): Same for condition 2.

Use F for M and X for A to get a ratio with no changes in density or velocity.

    F2/F1 = (ρ2 • X2 • V2)/(ρ1 • X1 • V1)

Because ρ2 = ρ1 and V2 = V1:

    F2 = F1 • X2/X1

517. The correct answer is (B).

In some cases, the best action is to hold the previous position, at least in the short term.

518. The correct answer is (A).

On loss of power, the MX contact opens and the motor will remain off until either the local or the DCS start is activated, or the motor starter pulls in, sealing in the circuit.

519. The correct answer is (D).

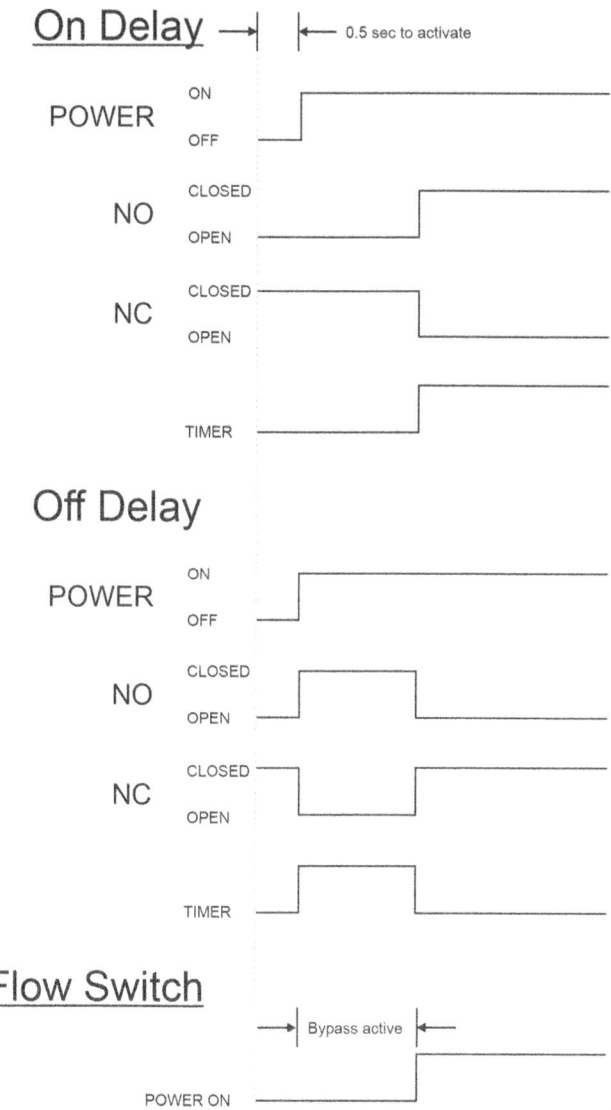

The start-up bypass could be either an on delay relay with a normally open contact or an off delay relay with a normally closed contact. The off delay relay with a normally closed contact is not fail-safe because the normally closed contact will bypass the low flow shutdown. The on delay relay typically requires a 0.5 s power pulse to start the timer, which in turn energizes the relay and the contacts change state (i.e., the normally open contact closes). Therefore, the on delay relay is a fail-safe design and the only solution.

520. The correct answer is (D).

A double-acting actuator must be given power to open and again to shut. A fail-safe device, if not already in safe mode, will go into safe mode on its own when it loses signal/control power.

521. The correct answer is (D).

This scheme blocks in the pressure on solenoid failure. If the solenoids are momentarily actuated, the scheme is also immune to loss of instrument air.

Distractors:

(A) Fails closed.

(B) Fails closed if the solenoid fails. It is unpredictable on loss of air, and it will not actuate in both directions.

(C) Almost works, but it may move when failed if the process exerts force on the valve.

522. The correct answer is (A).

523. The correct answer is (D).

This Boolean statement means that M is activated when A is closed OR if B is closed AND either C or D is closed.

524. The correct answer is (A).

To check the stability of a system having this characteristic function:

$$s^4 + 7s^3 + (19 + K)\, s^2 + (13 + 7K)\, s + 12K$$

Write the Routh array as follows:

s^4	1	$19 + K$	$12K$
s^3	7	$13 + 7K$	
s^2	a	b	

$$a = \frac{7(19 + K) - (13 + 7K)}{7}$$
$$= \frac{133 + 7K - 13 - 7K}{7}$$
$$= \frac{120}{7}$$

s^1	c	$b = 12K$
s^0	d	

$$c = \frac{a(13 + 7K) - 7b}{a}$$
$$= 13 + 7K - 7(12K)\left(\frac{7}{120}\right)$$
$$= 13 + 7K - 4.9K$$
$$= 13 + 2.1K$$

$$d = b = 12K$$

The first column will be positive when $c > 0$ and $d > 0$. The first condition requires $K > -6.2$; the second requires $K > 0$.

525. The correct answer is (D).

I   is false. The parameters are the period of the oscillation and the gain needed to produce the oscillation.

II   is true.

III is false. Three or four cycles are sufficient.

IV   is true.

526. The correct answer is (A).

Pure time delay requires a reduction in both proportional gain and integral action.

(See D. R. Coughanowr and L. B. Koppel, *Process Systems Analysis and Control*, McGraw-Hill, 1965, 312–314, where the Cohen–Coon tuning formulas are discussed. The equations show Kc varying inversely with dead time, which means that the integral action decreases with dead time.)

527. The correct answer is (C).

528. The correct answer is (D).

A room's temperature is to be maintained in spite of load disturbances. This question deals with the difference between tuning for set-point tracking or disturbance rejection.

529. The correct answer is (B).

The interface is at 0% when only oil is present:

$$(14 \text{ in} \bullet 0.82) - (14 \text{ in} \bullet 0.934) = -1.546 \text{ in wc}$$

The interface is at 100% when only water is present:

$$(14 \text{ in} \bullet 0.99) - (14 \text{ in} \bullet 0.934) = 0.784 \text{ in wc}$$

530. The correct answer is (B).

The output of timer "A" will not go to 1 until after 10 min. Because its output is negated by the *not* going into the 4-input AND gate, the result is that the motor is stopped.

531. The correct answer is (C), based on clause 1 of ANSI/ISA-61511-3-2018 /IEC 61511-3:2016.

532. The correct answer is (A), based on clause 13.1.1 of ANSI/ISA-61511-3-2018 /IEC 61511-3:2016.

533. The correct answer is (C), based on clauses 3.4 and 3.5 of ANSI/ISA-61511-3-2018 /IEC 61511-3:2016.

534. The correct answer is (D), 0.9988.

Solution: Probability of no failures = $(0.98)_3$ = 0.941192

Probability of one failure = $3 (0.02)(0.98)^2$ = <u>0.057624</u>

Probability of system operating = 0.998816

As a check: Probability of two failures = 3 $(0.02)^2$ (0.98) = 0.001176

Probability of three failures = $(0.02)^3$ = 0.000008

Probability of system being shut down = 0.001184

Total of all outcomes = 1.000000

535. The correct answer is (C).

This is Equation 14 from AGA Report 7 (1996), *Measurement of Gas by Turbine Meters.*

536. The correct answer is (B) according to FOUNDATION Fieldbus installation guide AG-181, revision 3.2.1, section 7.2.2.

537. The correct answer is (B).

Replacement in kind (exact model, calibration, connection, etc.) does not require that an MOC be conducted. Any change to a system that is not in-kind requires an evaluation of the potential hazards that may be introduced.

538. The correct answer is (B).

539. The correct answer is (D).

This is required by NFPA-85C, *Standard for the Prevention of Furnace Explosions/Implosions in Multiple Burner Boiler-Furnaces,* and precludes the selection of (A), (B), or (C). In addition, most PLC manufacturers recommend an external master relay to remove all power from field devices in an emergency.

540. The correct answer is (A).

Optical radiation can be absorbed by other surfaces and be a source of ignition. If the wavelength matches the absorption band of a flammable gas, then thermal ignition can occur. Light emissions come in many different wavelengths, from IR to UV. These waves can react with oxygen molecules in the area to generate an "oxidant" that could ignite. If a "laser beam" of optical radiation hits a potentially explosive gas, it can generate plasma or a shock wave—both of which are ignition sources.